A Theory of Epistemic Justification

PHILOSOPHICAL STUDIES SERIES

VOLUME 112

For further volumes:
http://www.springer.com/series/6459

Jarrett Leplin

A Theory of Epistemic
Justification

 Springer

Dr. Jarrett Leplin
5623 Brisbane Drive
Chapel Hill NC 27514
USA
jleplin@nc.rr.com

ISBN 978-1-4020-9566-5   e-ISBN 978-1-4020-9567-2

DOI 10.1007/978-1-4020-9567-2

Library of Congress Control Number: 2008941179

Printed on acid-free paper

9 8 7 6 5 4 3 2 1

springer.com

Preface

One goal of epistemology is to refute the skeptic. Another, with an equally distinguished if briefer pedigree, is to make sense of science as a knowledge-acquiring enterprise. The goals are incompatible, in that the latter presupposes that the skeptic is wrong. The incompatibility is not strict. One could have both goals, conditioning the latter upon success at the former. In fact, however, epistemologies aimed at the skeptic tend not to get anywhere near science. They've got all they can handle figuring out how we can know we have hands.

I come to epistemology from the philosophy of science, my original interest in which was epistemological. Philosophers of science are concerned with epistemic justification, but their question about it is how far it extends. They take justification to be unproblematic at the level of ordinary experience; their worries begin with the interpretation of experience as evidence for theory. They are interested in the scope of scientific knowledge. Having taken a position on this question (1997), arguing that justification extends to theoretical hypotheses, I came to wonder about the nature of justification generally. This is not a belated discovery of the skeptical problem or a reconsideration of what I took to be unproblematic. It is simply an interest in the possibility of locating epistemic advance in science within a broader understanding of the nature of epistemic justification. Now that I know that justification extends to theory, I am taking a step back and asking what justification is.

Approaching general epistemology in this way, I have found, makes a significant difference to intuitions about cases and to what one is willing to assume. My own stance is partially naturalistic, in that I take real historical situations in which science progresses epistemically as data to which epistemology should be responsive. An epistemological theory that denies the justifiedness of what history identifies as epistemic achievement faces a heavy burden. Philosophy must make peace not only with intuitions about possible cases, but also with the historical diagnosis of actual cases.

So I assume that my belief that I have hands is justified. But I also assume that Newton knew more than Galileo. I assume that Kepler's great insights were justified, although he himself could not distinguish their epistemic status from that of his massive metaphysical confusions. The sciences abound in reliable methods of learning about the world, and philosophical problems about reliability or justification or knowledge do not compromise the epistemic status of the results these methods deliver. I can only hope that epistemologists schooled in a different agenda that disputes what I assume will bear with me.

I wish to thank Ram Neta for criticism, and for the tact he exercised in alerting me to anticipations in extant literature of ideas I had fancied mine alone. Thanks also to the other participants in the 2004 Greensboro Symposium in epistemology, especially William Lycan, David Christensen, and Fred Dretske. Thanks to a scrupulous anonymous referee for Springer. If there were any errors remaining in my text, their resistance to recognition or correction would not be attributable to any failure of diligence or acuity in my critics.

I thank the National Endowment for the Humanities for a fellowship in support of my research.

Chapel Hill, NC, USA   Jarrett Leplin

Contents

Chapter 1
Introduction

1.1 Overview

Reliability theories of epistemic justification and knowledge originated in the late 1970s and flourished through the 1990s. Although still influential, reliabilist epistemology is widely thought to be seriously defective, and has been largely superceded by such (purported) rivals as evidentialism and virtue theory. This book takes exception to this development. I contend that reliabilism remains an important part of the true story of justification. I will develop a new reliability theory free from the burdens that discredited earlier theories. But the solutions my theory offers to counterexamples and objections raised against other theories are incidental to its motivation and development. I derive my theory from a particular way of understanding epistemic goals, and from the assumption that we have, in ordinary life and in the sciences, standards for and ways of investing credence that advance epistemic goals thus understood.

In this introductory chapter I will identify inadequacies in two classic versions of reliabilism to set constraints on a successful theory. That is, I will advance conditions of adequacy for the success of any theory of epistemic justification. Some of what I say may recall familiar problems, but my critical aim is not to disprove alternative positions so much as to motivate the new direction I will pursue. Then I will forecast the development and defense of my theory by delineating the contributions of ensuing chapters.

1.2 Goldman's Process Reliabilism

A theory of epistemic justification is reliabilist if it makes the justification of a belief a function of the reliability of the process or method by which the belief is formed or sustained. Alvin Goldman proposes two quite different accounts of reliable belief-formation, one for (noninferential perceptual) beliefs that are to count as knowledge

J. Leplin, *A Theory of Epistemic Justification*, Philosophical Studies Series 112,
DOI 10.1007/978-1-4020-9567-2_1, © Springer Science+Business Media B.V. 2009

(1976), and one for the justification of beliefs (1979).[1] Here I shall be concerned with the later account. The former way of understanding reliability is better represented by Robert Nozick's theory of knowledge (1981), and will be the subject of the next section. Both Goldman and Nozick use the term "reliable" in connection with both types of account.

In (1979), Goldman understands the reliability of a process of belief-formation as its tendency to produce beliefs that are true rather than false. This tendency may be either a high frequency of truths among the beliefs actually formed by the process, or a propensity or disposition of the process to deliver truths if the process is realized. But reasoning—if it is good, correct, deductively valid, reasoning by some standard—is surely justificatory; at least it would seem so pre-analytically. It is epistemically better to believe rationally, as a result of reasoning, than to believe by wishful thinking or indoctrination, for example. And reasoning is neutral with respect to the truth-values of the beliefs it delivers. It can be assumed to deliver truths only conditionally on the truth of its inputs.

But it will not do to make reasoning *from truths* the process whose reliability is to confer justification. For then one who just *happens* to believe truths will be justified in *any* beliefs he rationally infers from them. Indeed, a person with no justified beliefs could convert a good measure of his beliefs into justified beliefs by the application of trivial tautologies. So it must be reasoning *from justified beliefs* that is to be the reliable, and hence justificatory, belief-formation process. And this recursion is indeed how Goldman provides for the justificatory property of reasoning.[2]

Justification, however, attenuates through reasoning. Unless one's reasoning is logically conclusive, it introduces a risk of error additional to that already carried by one's premises. If justification attenuates to the point that conclusions, because of the risk they carry, are no longer justified, then Goldman's proposed explication of how reasoning justifies does not work. This problem arises for Goldman's theory even if the reasoning is deductively valid. For beliefs justified by Goldman's reliabilist standard collectively carry logical consequences that are certainly (self-evidently) false, and hence unjustified. Rather than gradual and cumulative, the "attenuation" here is immediate and complete. So, for example, beliefs formed by a process that has a high probability of delivering truths may be collectively inconsistent. We then get an outright, patent contradiction by logical deduction from beliefs that Goldman counts as justified. Of course, Goldman knows this; he acknowledges the lottery problem in a footnote. His defense is that it is only the "ordinary (naïve) conception of justifiedness" (1979, note 10) that his theory is intended to explicate, and lottery-type problems do not inform ordinary thinking. And this is how he leaves it.

---

[1] According to Goldman, the former account is properly taken to apply to justification as well as to knowledge, if we do not (as he says we should not) presuppose a Cartesian, foundationalist conception of justification.

[2] At least, this is the gist of it. More precisely, Goldman defines the conditional reliability of belief-dependent processes as their tendency to produce true beliefs with true beliefs as inputs. Then a belief inferred from justified beliefs is justified if produced by a conditionally reliable process.

Lotteries are very popular. There is a lot of ordinary thinking about them. I doubt that the millions who purchase lottery tickets in the hope of winning believe that the tickets they are about to purchase will lose. If they believe this, why make the purchase? If, as Goldman contends, they are justified in believing this by their own ordinary standards of justification, why don't they believe it? And if, despite their behavior, they do believe their ticket will lose, then don't they also believe that their neighbor's ticket will lose, that any given ticket will lose—beliefs that Goldman says their own standards justify? At what point short of paradox does justification surpass conviction? Is it simply a failure to reason that isolates ordinary thinking from paradox, or do people decline to exercise their purported entitlement to form the beliefs to reason from?

Of course, one might take belief to be a qualified state of mind, short of outright conviction. Ticket buyers likely are inclined to believe, are reasonably sure, are highly expectant that their tickets will lose. They hold out little hope of winning and would be astonished if they won. But by "belief" I mean and I think Goldman means a state of conviction altogether incompatible with doubt. That is what he is proposing to explain the justifiedness of. If it were something weaker than this, we would get a very different theory, a theory not of when a belief is justified but of how much confidence in a proposition one is justified in having. So I am not sure that public insensitivity to the implications of believing that tickets will lose is safely assumed. I suggest that what creates the lottery paradox is not the discovery of these implications, but the advent of a philosophical theory that would justify believing that tickets will lose. The solution is not to arrest reason, but to deny the theory. On my theory of justification, unlike Goldman's, lottery beliefs are unjustified.

Even if interest in lotteries is without effect on ordinary standards of justification, it is implausible to suppose that these standards countenance the justifiedness of outright, readily recognizable contradictions. Surely, recognition that a proposition unavoidably errs subverts its claim to justification. Even one who has not put the lottery paradox together will balk at the proposition that although each ticket loses some ticket wins, a proposition justifiedly believable on Goldman's theory. And if blatant contradictoriness did pass ordinary muster, I think the philosophical task would be not just to explain ordinary standards but also to rectify them. In any case, one constraint I will impose on a theory of epistemic justification is that it not permit the justification of recognized contradictions. Goldman's theory violates this constraint.

I will not try at this point to decide whether restrictions on permissible patterns of reasoning can fix this problem, for it is not the ultimate disposition of Goldman's approach to justification that concerns me.[3] My purpose is to motivate an approach

---

[3] Although he wants reasoning to transmit both justification and knowledge, Goldman will deny that knowledge, and by implication justification, are closed under known entailment (1986): one cannot know by inference propositions that one's evidence does not discriminate from relevant alternatives. This restriction is a response to skepticism, and I will argue in Chapter 7 that it fails. I note here that applying the restriction to justification would not prevent Goldman's theory from licensing contradictions. Everyone losing is certainly discriminable from any relevant alternatives in the lottery.

that need not restrict reasoning to avoid paradox. With respect to this purpose, I have another, more fundamental difference with Goldman. I do not think that a theory of epistemic justification is or can be an explanation of a pre-existing ordinary conception of justification. There is no such conception, or, if that is too strong, there is in ordinary language no coherent conception of distinctively epistemic justification.[4]

The ordinary idea of justification is all mixed up with ideas of entitlement, blamelessness, vindication, integrity, exoneration, rationalization, expectations, norms, compliance with rules—standards that apply most immediately and least problematically to conduct rather than belief. Indeed, it is something of an aphorism that belief is unconstrained; you may believe what you like but must play by the rules. Instead of descending into the complex of common practice and attempting to extract something distinctively epistemic out of it, I propose to fix on epistemic goals and ask what advances them. Although it is not without problems and qualifications, there does seem to me to be, in ordinary thought, the distinctively epistemic goal to believe what is true and not what is false. A second constraint, or condition of adequacy, to place on a theory of epistemic justification is that justification attaches only to what in some way advances this goal. It is at best unclear whether Goldman's theory can satisfy this condition, for it necessarily justifies false beliefs along with truth beliefs. On the interpretation I shall defend, Goldman's theory violates the condition.

1.3 Nozick's Subjunctive Reliabilism

Robert Nozick's (1981) theory of knowledge provides an alternative to Goldman's (1979) way of understanding reliability. Nozick requires a method of belief-formation that yields knowledge to be subjunctively sensitive to the truth-value of what is believed. His preferred term is "tracking", not "reliability", and his requirement applies to knowledge rather than justification. When Nozick comes to justification and uses the term "reliable", he adopts a reliabilist standard more like Goldman's: the reliability of a method is the high probability that beliefs it produces are true (1981, p. 265).[5] But that a method of forming beliefs tracks the truth-values of the beliefs formed would seem, intuitively, to be a very strong form of reliability. Nozick himself points this out (p. 265, note), and Goldman classifies tracking as a reliability theory (1986, p. 45). So describing subjunctive sensitivity to truth-value as "reliability" conforms to the usage of both Nozick and Goldman. And if knowledge requires epistemic justification, then Nozick's conditions for knowledge

---

[4] In his later (1986) Goldman says that "no unique conception of justifiedness is embraced by everyday thought or language" (pg. 58). But he thinks that purely semantic considerations support a core notion of justification as compliance with rules, and he says that this is essentially a deontic conception. I consider deontic justification in Chapter 2.

[5] And, as noted above, Goldman has proposed a subjunctive condition for (noninferential perceptual) knowledge, which he describes as a reliability condition (1976).

need to be at least as strong as any conditions necessary for justification.[6] So his conditions would need to ensure reliability, if justification requires reliability.

The reliabilist condition that I will propose for justification is superficially similar to the third of Nozick's conditions for knowledge. So it will be useful for me to compare his theory with mine, even though his is a theory of knowledge and mine is a theory of justification. Nozick's third condition for knowing P via method M is: if P were false and S were to use M to decide whether or not P, then S would not believe P (1981, p. 179). In so far as it is M's verdict that determines what S believes, S does not believe falsely. If other methods act in concert with M, or would be used were M not used, then whether S knows depends on how the methods compare in their influence on S. Nozick's proposals for measuring influence will not matter for my theory. Clearly, a method satisfying Nozick's condition deserves to be considered reliable. Trust in it and you cannot go wrong.

But neither may you go right. A method that never delivers a wrong verdict may deliver few verdicts; it may leave many propositions to which it is applied undecided. Nozick adds the further, fourth condition that (roughly) if P were true and S used M to decide about P then S would believe P. So, combining Nozick's conditions, deciding P by using M matches whether or not S believes P to the truth-value of P. This is tracking.

But why can't one know by a method that violates Nozick's further condition, and so fails to track? Why, when M *does* pronounce as to P, can't one know P by using M, provided only that the third, reliability condition is satisfied? A clock can tell me the hour without telling me the second. Eyesight identifies the buffalo, but not the mouse. There is no decision procedure for theoremhood, yet mathematical proofs prove. Nozick says (1981, pp. 684–685, note 23) that his further condition is satisfied in the latter case, and suggests that cases like the former can satisfy it also, via the expediency of making S's method not M but "believing M's answer" (if any). The suggestion, I take it, is that this is a method that cannot be used unless there is an answer. So if P were true and S were to use this method, there would have to be an answer for S, correctly, to believe.

I do not find Nozick's new, derivative method plausible as an account of how methods that are sometimes silent but ever reliable affect the formation of beliefs. It is not as though one first determines whether one's method delivers, and then, when it does, decides to believe *that*. One does not normally decide what to believe at all. The decision, if any, is what method to use to find something out. Espying, clocks, and mathematical argumentation are ways of finding things out. Once these things are found out, there is then no *further* method that says to believe what is found out. What is found out has *already* been learned.

One wonders, if the method of belief-formation is to believe M's answer, what M itself is a method of doing. M cannot be a method for determining whether M

---

[6] Of course, Nozick's conditions for knowledge will not be considered justificatory if one presupposes an internalist conception of justification. Essentially the same caveat as to the relation of knowledge to justification issued by Goldman (note 2 above) applies to Nozick.

delivers an answer. If M is a method of obtaining an answer, there must have been, independently, a question as a precondition of M's use. Yet inconclusive methods can deliver knowledge without any prior question, as when one knows by looking. In mathematical reasoning one can discover theorems by exploring implications, without the status of any particular mathematical proposition being in question. One can happen upon information in a way that makes this information knowledge, without there having been any point antecedently at issue.

As a method, what is vision? Its product is visual imagery, but vision is not (normally) a method for constructing visual imagery. As a method what it does is show what is there; its product is then belief. One uses clocks not to bring numbers to mind, but to tell time. Clearly, M is *already* itself a method of forming beliefs, albeit one that sometimes does not work. Methods do not always work, and this imperfection cannot be finessed by making the assumption of success constitutive of the method.

I suggest that Nozick has mixed up two distinct limitations on methods: they are not always usable; and, when used, they do not always work. Methods are not always usable because there are preconditions for their application. But that a method works is not a precondition for its application. Vision is not usable in the dark. Used in the light, it may not work for objects tiny, remote, or obscured. Nozick's derivative method will not satisfy my theory's requirements for epistemically probative methods.

I find Nozick's further, fourth condition, whether proposed for knowledge or for justification, unacceptable, and my theory will not involve it. Its disqualification of probative, inconclusive methods is but one objection. Another is that it is unrealistic. I doubt that Nozick's third and fourth conditions for knowledge are simultaneously satisfied by real methods, at least I doubt that the ordinary sorts of methods he discusses satisfy both conditions. Even scientific methods rarely satisfy them. It would take an unusually sophisticated and specialized scientific procedure heavily dependent upon background theory to satisfy both conditions. Normally, justificatory scientific procedures do not guarantee, nor even carry a high probability of delivering, results. Often it is just good, improbable fortune that we are able to realize conditions under which a consequence by which to judge a hypothesis is predictable. The required conditions could depend on technological advance or the unlikely cooperation of nature, as when we build a machine to replicate the early universe or hope for an alignment of galaxies to test gravitational lensing.

My root objection to the fourth condition is that a method of belief-formation that guarantees going right carries an unacceptable epistemic cost of also going wrong. A method bound to deliver the belief that P whenever P is true, may also deliver the belief that P when P is false. It reliably delivers beliefs only by being unreliable as to the truth of the beliefs it delivers. This result is avoidable only via disputatious restrictions on the quantifier "whenever". Nozick has to assume what subjunctive situation one would be in were P true, and what the situation would be were P false, such that in these situations M gets P right. I do not see that he has any principled basis for assumptions that will guarantee a successful M. Preferring safety to sorrow, I opt for condition three; epistemic goals are better served by methods reliably correct than by methods reliably productive.

My first point about understanding reliability via Nozick's third condition is not a criticism, just an observation. If your analytic objective is knowledge and you assume that what is known is true, then you are allowed to propose conditions that false beliefs cannot satisfy. By contrast, it must be possible for false beliefs to satisfy conditions for justification. This makes the theory of justification more complicated, in one respect, than the theory of knowledge, even if justification is a condition for knowledge. Even if a theory tells us correctly what knowledge is, and everything it says is knowledge is justified, the theory leaves us unable to distinguish justified from unjustified beliefs in general unless it *also* tells us what justification is. So one major difference between my version of reliability and Nozick's condition will be that I provide a way for reliably produced beliefs to be false. Of course, my way will be different from Goldman's, in view of the constraints I have already endorsed.

Much critical reaction to Nozick concerns the identification and individuation of methods. Suppose that M_1 satisfies Nozick's reliability condition but M_2 does not. What makes it the case that S uses M_1 rather than M_2? I am not asking how we *tell* which method S uses. We might be unable to tell, and so unable to tell whether S knows, and yet Nozick could be right about what it is to know. The question, rather, is what it is to use a method, and to use one method in particular. It cannot be a matter of what one recognizes oneself to be doing or of what one is aware of doing. It cannot require deliberation, premeditation, or the execution of a plan. The problem is not just that justification and knowledge are possible unawares. More importantly, one's awareness could be misdirected. Could one not be mistaken about how one's belief is formed? Could one not think that one is using M_1 while in fact one is using M_2? What if one mistakes a legitimate authority for a psychic and, because of this error, accepts his testimony? If one can mistake one's method, then what one is aware of doing does not decide what method one uses.

Nozick does not tell us how to decide. He acknowledges difficulties that the dependence of epistemic status on method raises, but says little in response to them. He says that when a belief is based on experience, it is only the subjective quality of the experience, not its object, that affects one's method (pp. 184). Seeing and seeming to see are the same method. Nozick does not comment as to whether one might be unaware of, or mistaken about, the subjective quality of one's experience. But certainly he is not proposing subjective experience alone as the determinant of method. He does not think that his reference to the subjective quality of experience settles questions of method.

He also says, "which method a person actually is using will depend on which general disposition to acquire beliefs (extending to other situations) he actually is exercising" (p. 185). But how is the exercise of a disposition any more determinate than the use of a method? It sounds like we are to rely on long-term behavior, and attribute the method that best fits with the subject's otherwise evidenced proclivities for investing credence. This is suggestive, nothing more. The immediate concern is that a method could be used only once; possibly only one, or a limited, aberrant few beliefs are justified. Possibly knowledge or justification is uncharacteristic for this subject. Of course, one need not admit these possibilities, but then there is the further argumentative burden of excluding them, a burden that Nozick does not accept. Another major difference between my theory and Nozick's is that mine develops conceptual resources to delineate methods.

My principal departure from Nozick concerns epistemic closure, the transmission of epistemic properties through reasoning. Nozick denies that knowledge is closed under even some of the most self-evident, logically conclusive forms of inference. We cannot just by inference know instances of universal generalizations that we know, nor conjuncts of conjunctions that we know, for example. The reason is that the property of subjunctive sensitivity to truth-value—tracking—which Nozick requires of a belief that is known, does not in general hold of a belief's logical entailments.[7] The condition of subjunctive sensitivity to truth that I will count sufficient (almost; there will be more to it) for epistemic justification, my reliability property, is satisfied by beliefs entailed by beliefs that satisfy it, supposing the entailment to be the basis for believing them. On my theory, epistemic justification will be closed under any truth-preserving inference.[8]

Nozick proposes a restriction on closure that identifies when inference transmits knowledge: the belief from which one infers must track the truth of the belief inferred. If the conclusion were false one wouldn't believe the premises (in the way that one does come to believe them that enables one to know them). This is necessary, because if one did believe the premises there would be nothing to stop one from coming to believe the conclusion, which is inferable from them. But knowledge requires a situation in which a belief is false to be one in which it would not be believed. Also, if the conclusion were true one would believe the premises, so that one could come by inference from them to believe the conclusion. One knows the conclusion only if were it true one would believe it (although "could" is all we get here; one needn't infer). On my theory of justification, the first form of subjunctive sensitivity will be automatic to the extent that it is needed, and the second will be unnecessary. No restriction on closure is required. This is not merely a result

---

[7] There is, incidentally, a curious asymmetry in Nozick's analysis of how conjuncts of known conjunctions can be unknown. Suppose that I know, and so track, P but not Q. And suppose also that the falsity of $P\&Q$ would have to result from the falsity of P; it would not result from the falsity of Q because it wouldn't be Q that was false. (The close $\sim(P\&Q)$ worlds are $\sim P\&Q$ worlds.) Then I can track, and so know, $P\&Q$, although I do not track and so do not know Q. For if $P\&Q$ were false P would be false, and as I track P I would not then believe P. Not believing P I would not believe $P\&Q$, as required to track $P\&Q$. But why does Nozick assume that if I do not believe P then I do not believe $P\&Q$? If I can know $P\&Q$ without knowing Q, why can't I believe $P\&Q$ without believing P? Perhaps we can begin to make sense of this idea by considering cases in which people intuitively rank conjunctions higher in probability than one of their conjuncts. One may more readily believe that a politician will not seek reelection but will receive an administrative appointment, than believe simply that he will not seek reelection. If we are going to take seriously so bizarre a result as Nozick endorses about knowledge, we should be willing at least to reconsider corresponding assumptions about belief.

[8] Sherrilyn Roush (2005) also endorses closure within a tracking format. However, she simply adds closure to her theory as an independent condition. My closure condition is a consequence of my theory. I provide a theoretical basis for closure, rather than just stipulating it.

desirable if you can get it. I shall argue that truth-preserving inference must transmit justification, and I adopt the closure of justification under truth-preserving inference as a condition of adequacy for a theory of justification.

Giving closure the status of a constraint on theory raises the general question of whether it is necessarily beneficial epistemically. Granting that closure answers to epistemic desiderata, is it necessarily beneficial *on balance*? What if our inferential capacities were structured so that along with inferring consequences of justified beliefs we could not help but believe a lot of other stuff as well? The closure principle my theory prescribes obviates this problem. It justifies only beliefs entailed by justified beliefs, and it does not require or ordain the exercise of inferential capacities at all. Closure does not tell us to infer. It only tells us that if we do infer, then whatever damage is done we at least get justification for inferred consequences of the justified beliefs we infer from.

My theory is about justification, which is, at most, a necessary condition for knowledge. Closure could hold for justification without holding for knowledge, so that my condition of adequacy does not automatically set me at odds with Nozick.[9] Although I do, additionally, think that the reasons I will give for closure of justification apply to knowledge, my purpose here is not to dispute Nozick's position on knowledge but only to mark a significant contrast with the approach I will take. Nozick does not merely deny closure; he celebrates its failure. He counts this consequence of his analysis of knowledge a virtue. He thinks that an unrecognized, implicit but mistaken assumption of closure is responsible both for sweeping skeptical positions and for facile dismissals of skepticism. Denying closure is his key to salvaging ordinary knowledge while giving the skeptic his due. He treats nonclosure as a *discovery* about knowledge that credits his theory.

By contrast, I would regard denying closure, were this necessary, as a sacrifice to lament. Nonclosure, for me, is a refutation of any theory that requires it. Nozick wants to explain how knowledge is possible. Well, one way it is possible is through reasoning. Knowledge is obtained by applying principles of logical inference to what is already known. Inference from what one knows is a principal way of knowing. By not letting us get knowledge this way without further ado, Nozick implausibly limits its scope.

In making a virtue of sacrifice, Nozick appears audacious and extreme. But he is really just facing up to consequences already looming, although for different reasons, in other positions. Goldman, remember, does not explain how to extend justification through reasoning without paradox. And he (1986) agrees with Nozick that knowledge must violate closure to avoid skepticism. Peter Klein (1981) thinks that justification is closed under known entailment, but not under conjunction. He thereby he avoids the lottery paradox, but, I shall argue, unacceptably restricts the range of inference capable of transmitting justification. Neither of their positions satisfies my condition of adequacy.

---

[9] But in fact with respect to justification Nozick is in the same boat as Goldman, and will have to violate closure to avoid paradox.

My difference with Nozick over closure reflects a difference in attitude toward skepticism. Nozick argues that we can have ordinary knowledge despite not knowing (and despite being unable to know) that skeptical scenarios which falsify ordinary beliefs do not obtain. I think that this argument both underestimates and overestimates the skeptic's position.

Skepticism is underestimated because Nozick's defense of ordinary knowledge presupposes that skeptical scenarios do not obtain. If one did, then the situation that would obtain instead of what one truly believes, were one's belief false, might not be detectably different. And then one's belief does not track truth-value and so fails to be knowledge. Skepticism is overestimated because there is a way we can justifiedly believe (and I think know) that skeptical scenarios do not obtain, namely by truth-preserving inference from what we already justifiedly believe (or know). Like Nozick, I do not have the objective of refuting skepticism, but I do take a uniform position against it. I do not think Nozick can have ordinary knowledge without a uniform position on skepticism. He does not know the facts about his environment if he cannot know that he is not in a skeptical environment.

But *does* Nozick have ordinary knowledge, or merely its possibility? If we do not know, or are not at least justified in believing, that we have knowledge, or even that ordinary beliefs are justified, then although the skeptic could be wrong we are in no position to disagree with him. This is still a skeptical result, as Nozick himself acknowledges; his general characterization of skeptical argument (1981, p. 197) counts an impasse as a skeptical outcome. Skepticism is not a thesis about the existence or nature of ordinary objects of perception, but an epistemological thesis about knowledge or justification. To deny skepticism, it is not enough to claim that ordinary beliefs are true; we must claim to know or justifiedly believe them. Does Nozick claim this?

Although his official purpose is only to explain how knowledge is possible (1981, p. 167), in many places he claims it is actual. He says repeatedly not just that many things are true but that that we know them to be true. He says the skeptic is wrong to deny that we know these things (1981, p. 209). This seems to intend an error of fact, not merely of excess. He says we know the actual world to be such that if what we ordinarily know were false, this would not be because any skeptical scenario is true. On the contrary, if what we ordinarily know were false, every skeptical scenario would be false as well (p. 200). Compatibly with skeptical possibilities, how can he know this? The answer that what would make this false is not a skeptical scenario but a situation in which he would not believe this generates the same question. Unable, by his own lights, to know that skeptical scenarios do not obtain, he is obliged to answer this question. I am not; on my theory the question does not arise.

1.4 Plan

The next four chapters develop and explain my theory of epistemic justification. The four chapters following these defend the theory against objections rooted in rival theories. I then devote a chapter to problems that require refinement and elaboration of the theory. The final chapter examines the coherence of the theory under self-application. That is an overview. The remainder of this section gives a more detailed picture of the structure of the work.

Chapter 2 fixes the target for a theory of epistemic justification by presenting a preliminary explication of the concept. A preliminary explication is needed because the concept is not perspicuous in ordinary language. I explain and defend an externalist view of epistemic justification as justification that advances the distinctively epistemic goal of believing truths without believing falsehoods. Epistemically justified beliefs are beliefs formed or sustained in ways that are truth-conducive. I criticize the rival internalist view that justification can depend only on grounds that the believer is in a position to appreciate or recognize, or can reasonably have been expected to be in a position to appreciate or recognize. Internalism respects the intuition that it is unfair to deny justification for reasons that could not (reasonably be expected to) influence the believer. I construct an argument to show that internalism lacks the resources to ensure this form of fairness. It could be unfair to expect the believer to assess beliefs correctly by the standards that it is fair to expect him to apply. The internalist intuition might as well be violated, for it cannot consistently be implemented.

Because it is part of the epistemic goal not to believe (any) falsehoods, justification is not to be awarded on any basis that generates falsehoods unavoidably. So some otherwise plausible bases for belief are not justificatory. In particular, that a proposition is highly probable does not justify believing it. This result immediately distinguishes my theory from Goldman's version of reliability. Truth-conduciveness, as I understand it, is an unusually strict standard for justification. But unless it is unachievable, its strictness should not disqualify it. I require the *possibility*, though certainly not the likelihood or feasibility, of investing credence in ways that avoid error altogether.

Chapter 3 analyzes the truth-conduciveness of ways of investing credence in terms of their reliability. I follow Nozick in giving a subjunctive conditional account of reliability, but incorporate a condition of normalcy: a method of forming beliefs is reliable if it would not form false beliefs under normal conditions. Supposing a true belief to be formed by a reliable method under normal conditions, were this belief to have been false the method would not have produced it under those conditions.

Normal conditions are conditions characteristic of occasions and environments in which the method is usable, whether or not the method is then (or there) used. So not only is it possible for beliefs produced by a reliable method to be false, it is even possible for such beliefs to be predominantly or exclusively false. For, a reliable method could be used predominantly, even exclusively, under conditions abnormal for it. Thus, reliability has nothing to do with the frequency with which truth is achieved. I give two criteria for assessing normalcy, which, I argue, come together to identify a single, coherent notion. Normal conditions include both conditions the believer presupposes in using a method and conditions that are prerequisite to the method's use.

Reliably formed beliefs are not simply beliefs delivered by a reliable method; it is further required that the method be used intentionally. This does not imply that reliable belief-formation is deliberative or voluntary. The test of one's intentions is what one does and would do under varying conditions. The evidence tests provide underdetermines one's intentions, but rivalry among hypotheses as to what method one intends is adjudicable on the basis of further testing.

Although what one does in forming a belief may be described at varying levels of generality, and so described varies in reliability, what one does intentionally does not generally admit of this variation. Thus, the problem of generality that has bedeviled other reliability theories does not automatically apply to reliably formed belief, as I understand it. One may, however, use different methods intentionally in arriving at a single belief, and these methods may differ in reliability. The belief then counts as reliably formed if any of these methods is reliable. My theory requires no measure of relative weight for methods.

A method of belief-formation could be inerrant without justifying the beliefs it forms. In the generic case, the subject is made to believe only what is true, but could as readily have been made to believe anything else. There is no danger that such beliefs will turn out justified on my theory, for they are not reliably formed. The subject does nothing intentionally to form them, if indeed the subject can be said to form them at all. Inerrancy is not reliability.

Chapter 4 uses the notion of reliably formed belief to introduce conditions for justification. These conditions form the core of a theory of epistemic justification, to be developed and refined into the full theory stated in Chapter 10. At the core is a distinction between the justification of belief and the justification of believing. Essentially, beliefs are justified if reliably formed; they are believed justifiedly if the believer has good reason to believe them to be reliably formed (whether or not he does believe this of them). This distinction disarms a good deal of (though not all) internalist opposition to the externalism of my theory. The distinction provides a respect in which a person unable to tell that his standards or methods of investing credence fail to be truth-conducive may nevertheless believe justifiedly. The independence of justified belief from justified believing does not redeem internalism, however. It is insufficient for justified believing that one's reason to believe one's method of belief-formation reliable be good by internalized standards. One's reason must really be good, and not just taken to be good, however blamelessly.

Conversely, belief-formation that is in fact reliable despite being unworthy of trust, or even deserving of distrust, does not justify the believer but at most his beliefs. And not even the beliefs are justified unless the method of belief-formation is used intentionally. So, for example, there is no danger on my theory that a clairvoyant will believe justifiedly, nor even that his beliefs will be justified. Clairvoyance is shown in Chapter 9 not to be, in principle, a source of belief at all. But already in Chapter 4 the intentionality requirement for reliable belief-formation diffuses such cases.

The possibility of blameless misassessment of the quality of reasons separates rationality from justification. Rationality does answer to internalist standards, and so reliabilism is not a theory of rationality. Chapter 4 proposes an explanationist account of what makes a reason to believe a method of belief-formation reliable a good reason. Either the reason or the reliability of the method must be assumed to explain the other. My theory does not require a general analysis of good reasons, nor would such an analysis supplant reliabilism. The justification of beliefs does not require reasons.

Beyond its accommodation of (some) internalist intuitions, the division of justified belief from justified believing has a number of important applications. It enables one justifiedly to believe mutually inconsistent propositions, provided that they are not justifiedly believed to be inconsistent. Although mutually inconsistent beliefs cannot all be justified, they can all be justifiedly believed. Thus, undiscovered, or undiscoverable, inconsistency does not defeat justification (Chapter 4). The division further shows how reasoning transmits justification, producing different versions of epistemic closure (Chapter 5). It explains how a contextual change that raises the stakes for being right can affect what one believes justifiedly, without making the property of being justified contextual (Chapter 7). It distinguishes conditions in which believing correctly by luck pre-empts justification, from conditions in which luck is epistemically innocuous (Chapter 8). It frees justification from adventitious constraints that virtue theories impose to disqualify beliefs formed by methods whose reliability is undetectably restricted (Chapter 9). It corrects misrepresentations of the social dimension of justification (Chapter 9). It protects the closure principle for justification against the objection that inference from reliably formed beliefs does not justify the belief that these beliefs were reliably formed (Chapter 9). And the division is used to defend the necessity of reliability for justification (Chapter 10).

Thus, in my theory, the independence of justified belief from believing justifiedly resolves many important problems that arise for reliabilism. The analysis of reliable belief-formation in Chapter 3 establishes this independence while maintaining justification's essential connection to truth-conduciveness.

Under the conditions for justification advanced in Chapter 4, beliefs formed by truth-preserving inference from reliably formed beliefs are justified. Hence, on my theory, justification is closed under truth-preserving inference. Chapter 5 elaborates and defends this result. In particular, supposed presuppositions of justification do not create exceptions to closure. That the truth of what one truth-preservingly infers from justified beliefs is a precondition for the justification of these beliefs does not prevent the inference from transmitting justification.

The closure of justification under truth-preserving inference is shown to require the closure of justification under conjunction. Accordingly, any conjunction of justified beliefs is justified. This result dissociates justification from probability, which decreases through conjunction. Justification does come in degrees, but degrees of justification do not behave like probabilities. (Numerical measures of degree of justification would not obey the axioms of probability.) Instead, the degree of justification of a conjunction equals that of its least justified conjunct. As a result, justification does not diminish through inference.

Chapter 5 argues that scientific reasoning requires my closure principles. Bayesian confirmation would seem to offer an alternative interpretation of science without closure. But to make subjective probabilities converge, and thereby to provide for the objectivity of science, Bayesianism requires that the evidence on which probability assignments to theories are conditionalized be believed. Hence, Bayesianism cannot replace a theory of the justification of belief.

Like Bayesianism, the default-and-challenge model of justification limits epistemic assessment to the revision of belief. And this model is similarly incomplete. That beliefs ordinarily carry a presumption of justification in practice, so that the evaluative burden falls not on the believer but only on his detractor, does not imply that beliefs do not need justification and so does not obviate the need for a theory of their justification.

Chapter 6 contends with paradoxical consequences of closure. Because justification is closed under conjunction, allowing justified inconsistencies does not resolve the lottery paradox unless contradictions are also justified, which contravenes one of my conditions of adequacy. Chapter 6 argues for a different resolution of the lottery. The dissociation of justification from probability implies that beliefs to the effect that individual lottery tickets lose are not justified by their high probability. This result does not threaten the justification of ordinary fallible beliefs, because the fallibility of a belief cannot be identified with the possession by its negation of positive probability. In general, ordinary beliefs do not have probabilities. Where they do have probabilities, these are not the basis of their justification.

There are special features of the lottery, like uniformity of probability assignments, and one might suppose that high probability suffices for justification conditionally on the absence of such features. However, Chapter 6 proves that if high probability justifies any belief then a contradictory belief is also justified. Closure then produces a justified contradiction, in violation of a condition of adequacy. It is common to concede that individual lottery beliefs are not known, but nevertheless to contend that they are justified. In fact, the reasons to deny knowledge apply equally against granting justification.

Closure also raises the paradox of the preface, in which a conjunction of justified beliefs, although not contradictory, is supposed to be unjustified because its justification is defeated by independent information. Chapter 6 argues that the kind of information in question cannot defeat the justification of the conjunction without defeating that of the conjuncts, so that the closure of justification under conjunction is not violated. In particular, second-order evidence against the reliability of belief-systems is at best inconclusive. In general, information that defeats the justification of a conjunction without bearing differentially on its conjuncts defeats the justifications of all the conjuncts.

Chapter 7 defends my theory against supposed skeptical consequences of closure. The purpose of this defense is not to refute skepticism, but to demonstrate that opposition to skepticism is not a reason to prefer theories that deny or qualify closure. Any threat that the possibility of skeptical scenarios poses to the justification of an ordinary belief incompatible with them is independent of my closure principles for justification. So there is no advantage against skepticism in denying closure. I consider a number of anti-skeptical strategies that depend on denying or restricting closure and show that none works. Their common failure is to beg the question against skepticism by presupposing that skeptical scenarios do not hold. For this presupposition is a condition for identifying the range of the alternatives to the truth of one's belief that one's justification of it is required to rule out.

If skeptical scenarios are not refutable, then it is convenient for the justification of ordinary belief not to depend on refuting them. According to a contextualist view of justification, what is at stake in being right determines what alternatives to the truth of one's belief must be refutable. As the stakes rise and fall justification appears and disappears, because the burden of refutation for unrefuted alternatives shifts. Other versions of contextualism require the truth-conditions for being justified, or the property of justification being attributed, or the meaning of "justified" (if this is different) to change with context. In any version, contextualism threatens my theory with suspensions of closure. A belief justified in a context will carry consequences that are unjustified and do not need to be justified within this context. Contextualism is supposed to be consistent with closure, because in contexts that make skeptical possibilities relevant ordinary beliefs are unjustified. Closure is not violated if there is no justification to transmit. However, it is not open to me to rescue closure in this way. In my theory, skeptical contexts do not defeat ordinary justification, so contextualism is unacceptable.

The core of my argument against it is that belief is characteristically stable across variations in the extent to which error is disadvantageous. This stability belies the view that justification is contextual. What changes with context is the rationality of acting on belief, not the justification of belief. For if justification were contextual, so would be belief itself. It is implausible to suppose belief robust against recognition of changes in its epistemic status.

Attempts to construct a standard of relevance for the alternatives that one's justification of a belief must rule out presume that only relevant alternatives can affect justification. But it is not clear that relevance, even if we could decide on a standard for it, is the right concept to measure justifiedness. By any standard, one's epistemic entitlement to a belief could depend on pushing investigation into the irrelevant, and alternatives that do qualify as relevant could innocuously be left uninvestigated. It seems that the one constraint that epistemologists universally respect in proposing standards of relevance is that they rule skeptical scenarios irrelevant. Maybe there is no more to the relevance of an alternative than that it not support skepticism. But to fix the boundary of the relevant at whatever point alternatives render the falsity of one's favored belief undetectable amounts to saying that justification requires doing as much and as little as can be done. Clearly, justification sometimes requires more than this, and sometimes less.

The semantics of possible worlds is frequently used to explicate the range of conditions within which one's justification for a belief must be able to discriminate the truth of the belief from alternatives. The worlds in which whether one believes is subjunctively sensitive to truth are restricted by some proximity or similarity condition. On my view, however, possible worlds are conceptually derivative with respect to subjunctive conditions, and do not explicate them. Chapter 7 argues that the legitimacy of subjunctive conditions in analysis does not depend on supplying them with a categorical base.

Although Chapter 7 does not have the burden of refuting skepticism, it appeals to plausibility as an adequate response to skepticism. No argument for skepticism can be as plausible as the ordinary beliefs whose justification skepticism denies. Unlike the other anti-skeptical strategies discussed, this appeal to plausibility does not beg the question against skepticism.

Chapter 8 explains why tracking is not required for justification. The argument against tracking applies to knowledge as well. I construct counterexamples to Nozick's fourth condition, whether proposed as a condition for knowledge or for justification. One can know or believe justifiedly through the good luck of being insulated against false or misleading but justifiedly believable defeaters of one's belief. One can know what one could easily have failed to believe under conditions that hold one's method of belief-formation constant. Both knowledge and ignorance can be epistemically lucky, in that justificatory belief-formation can depend on either. Although being right by luck generally contrasts with being justified, as it does with knowing, some of the ways that luck affects belief-formation are epistemically innocuous. Tracking gets knowledge and justification wrong because it is excessive and indiscriminate in its disqualification of beliefs right by luck.

My reliability condition for belief-formation requires counterfactual sensitivity to the falsity of beliefs it justifies. The contrapositive of sensitivity, safety, is often imposed on knowledge to prevent beliefs from being known by luck. Knowledge does not depend on luck because, according to safety, known beliefs remain true across a range of proximous counterfactual situations in which they continue to be believed. Chapter 8 compares sensitivity with safety. I argue that what reason there is to deny that subjunctive conditionals contrapose (are equivalent to their contrapositives) is not reason to distinguish safety from sensitivity as an importantly different condition for justification or knowledge. In particular, switching from sensitivity to safety gains no ground against skepticism. In general, skepticism aside, counterfactual situations in which an insensitive belief would be held though false include situations in which safety requires holding a belief to ensure its truth. As a consequence, where sensitivity fails so does safety.

Problems for reliability that I address by incorporating conditions of normalcy and intentionality into reliable belief-formation have driven other philosophers to abandon reliability in favor of virtue theories. The argumentative burden of Chapter 9 is to discourage this alternative. Chapter 9 opposes virtue theories that require, for the justification of a belief P, that one possess an intellectual faculty or capacity for correctly judging the truth of propositions across a range to which P belongs. I explain how justificatory, reliable belief-formation differs from the exercise of such a faculty, and criticize as misdirected the additional constraints that virtue theories place upon justification.

In particular, coherence conditions are properly directed at the attribution of belief itself, not at its justification. As a corollary, the clairvoyance-type cases that have troubled reliabilism, and externalist epistemological theories generally, are not coherently describable. No faculty of clairvoyance can coherently be imagined to impart beliefs incongruous with expectations, experience, and background beliefs, because such incongruousness violates a coherence condition for the attribution of beliefs. One cannot believe through a faculty of clairvoyance which one has reason to distrust or to believe impossible. Moreover, even if clairvoyance were somehow freed of problems of coherence and incongruity, it would still pose no credible counterexample to reliabilism. I argue independently that belief requires the agency of the believer. It follows that clairvoyance cannot deliver beliefs.

In shifting the focus of justification from the particular belief justified to the range of one's evaluative competence, virtue theory errs in opposite directions. It fails to block the justification of beliefs formed in epistemically irresponsible ways, if these beliefs are subsumed by one's range of competence and could have been formed justifiedly. For virtue theory cannot block such justification without abandoning its purported improvements over simple reliability. And virtue theory does block justification where it shouldn't. For the property of membership in a range of propositions that fall under one's intellectual competence is not in general preserved across known entailments. Virtue theory therefore violates my condition that justification be closed under truth-preserving inference. This violation applies both to justified belief and to believing justifiedly.

Chapter 10 raises new problems for my theory. There are apparent counterexamples both to the justifiedness of reliably formed beliefs and to the reliable formation of justified beliefs. These examples occasion a number of refinements to the theory as presented in Chapters 4 and 5. One is that the reliability of a method is sufficient to justify a belief only on the condition that the method remains applicable under the counterfactual supposition that the belief is false. Despite this requirement, counterfactual applicability is not a necessary condition for reliable methods to be justificatory.

The purported counterexamples in Chapter 10 press the question of the scope and limits of my theory. The most defensible interpretation of these is that reliabilism provides only sufficient conditions for justification. Other sources of justification are possible, subject to the constraint that irreducibly different sources of justification must all serve the epistemic goal. There is no reason a priori to expect there to be only one way to serve this goal.

Chapter 4 provides a theoretical basis for degrees of the justification of believing, since some good reasons for believing a belief to be reliably formed are better than others. Chapter 10 provides such a basis for the justification of beliefs. This basis depends partly on the notion of the domain of a method's reliability. Chapter 10 explains this notion and considers the possibility that the reliability of a method varies across its domain.

Chapter 11 examines the internal coherence of the theory and compares it in this respect to other epistemologies. Consistency under self-application—self-referential adequacy—challenges any philosophical theory that lays down conditions for an acceptable theory. Are theories of knowledge or justification knowable or justified according to the standards they themselves impose? Need they be? I contend that my theory fares no worse on this account than others, and fares better than most. But self-referential adequacy is at most a necessary condition of a theory's acceptability; a false theory can easily satisfy this condition. In case my defense of my theory in this book falls short of what the theory itself requires for justified belief, our epistemic attitude toward the theory should be something less than belief. For if the theory is true, something more than or different from my defense of it would then be needed to justify believing it.

As a fallback position, I propose a version of instrumentalism for epistemological theories, and for other philosophical theories that propose analyses of concepts and test these analyses against intuition, that enables us to help ourselves to the useful applications of theories without committing ourselves to the truth of these theories. Even if one is not justified in believing an epistemological theory, one may be justified in believing the theory's verdict as to the justification of beliefs in specific cases—including problematic or disputatious cases—to which the theory is applied. Instrumentalism was designed for science, but it fits much of philosophy better.

Chapter 2
Truth-Conduciveness

2.1 The Epistemic Goal

The theory of epistemic justification I shall advance is based on an assumption: epistemic justification is justification that promotes the epistemic goal of believing truths without believing falsehoods. This chapter explains why an assumption is needed, motivates my choice of what assumption to make, and clarifies what is being assumed.

It is, to begin with, problematic to impute goals in epistemology, regardless of what one takes them to be. Whose goal is it, and how do we know? I do not attribute what I take to be the epistemic goal to individual cognizers, nor claim to read it off of epistemic practice, say as the best explanation of what cognitive agents do. What agents do underdetermines their goals. Real agents are many things besides cognizers, and I would not know how to identify the cognitive part of practice without an epistemic goal already in mind. I simply assume that believing truly has intrinsic value, and that this value is codified in a goal that is distinctively epistemic, as against, say, moral, aesthetic, or pragmatic.

I shall refer to this goal as "the epistemic goal", although other goals may also be regarded as distinctively epistemic. Knowledge, conformity of belief to evidence, fulfillment of epistemic duties, and consistency are natural candidates. I think that believing truly outweighs such goals, but this is not part of my assumption. The priority of truth over some goals is more properly argued than assumed, and its relation to others need not matter, because a theory of justification pursuant to them need not conflict with the theory I shall develop. Evidence, knowledge, and deontology will be considered in due course. Here I just note that the goal of consistency strikes me as wrong-headed. Surely it is better to be right than consistent. If consistency requires error, then inconsistency is the virtue. Better to change one's mind than to compound error. We promote consistency as such only where we think the truth elusive in principle.

Of course, the intrinsic value of true belief is conditional on belief. That is, believing truly has intrinsic value given that one is to believe at all. It can certainly be better not to believe, better to protect oneself against the intrusion of matters that do not deserve one's attention, better to remain ignorant to achieve the pleasure of

J. Leplin, *A Theory of Epistemic Justification*, Philosophical Studies Series 112, DOI 10.1007/978-1-4020-9567-2_2, © Springer Science+Business Media B.V. 2009

surprise or avoid the anguish of disappointment, better to be out of the loop and so unaccountable. Surely it is no one's goal to believe truly as such; one wants interesting or important truths subject to certain conditions. I have not much of interest or importance to say about what makes beliefs interesting or important. This depends on the much more specialized goals of the individual, and may not be generalizable. People may certainly believe in absolute standards of interest or importance; but as what they believe those standards to be varies, so do their epistemic goals.

Nevertheless, beliefs formed in an epistemically propitious way, beliefs suitably grounded (let us say), must be eligible for justified status, whether they serve individual epistemic goals or not. The conditions for justification, whatever they are, cannot include the service of one's individual goals. For it does not render a belief unjustified that the believer would just as soon not have formed it, or that forming it conflicts with his individual interests. Justification serves the general interest of believing truly. What one does to serve this interest is justificatory, even if this is not one's interest.[1]

In what respect, then, is the epistemic goal a "goal" if no one has it? I suggest that it is implicit in the following way. Ignorance may be preferable to true belief, and ignorance includes both no belief and false belief. It is the former only that one prefers to true belief. The later may be preferred to the former, but not to true belief. That is, a case in which false belief is preferable to no belief is not a case in which no belief is preferable to true belief. Believing falsely is preferable to not believing when it is the state of believing itself that has instrumental value, not its truth. For example, one often hears that the electorate admires conviction regardless of its content. Thus is explained the election of politicians whose positions are unpopular. It could be expedient for a politician to have beliefs, even if they are false. The truth of what the politician believes is incidental to his interests.

In such cases, I suggest that falsity is tolerated for the sake of believing. One can prefer believing falsely over not believing, but the falsity of the belief is not an attraction; it is a compensated disadvantage. If one prefers believing falsely to not believing, this is because of the misfortune that what one needs to believe is false, or that one's credence is misdirected. Of course, one can prefer that the truth be different, but, given what the truth is, this is not a preference to be mistaken. A preference to be mistaken is a preference that the truth be otherwise, not that one believe otherwise than what the truth is. To prefer false belief to true belief would require a special pragmatic motivation that trumps one's epistemic interests. One is not indifferent to falsity.[2]

---

[1] Thomas Kelly (2003) would like half of this. He argues that people can have justified beliefs that do not advance any epistemic goal of their own, but he does not like the idea of a general epistemic goal attributable to no one. His point against a general goal, though, is only that it does not redeem instrumental rationality. I can grant him that. Instrumental rationality extends only to methods, not to the beliefs they produce.

[2] Kelly (2003) says that with respect to matters in which one takes no interest, one has no cognitive goal better served by true than by false beliefs. He infers that in such matters, one has no preference for true over false beliefs. This inference is a non sequitur. Where one has no cognitive goal at all,

So what I wish to propose a theory of is that which advances an implicit preference for true belief. It is reasonable to ask why we should expect epistemic justification, so understood, to admit of theorizing. My assumption that justification advances the epistemic goal suggests the answer that the status of being epistemically justified has nonepistemic truth conditions. It is an empirical matter what in fact advances the epistemic goal. A correct theory about this would, according to the assumption, tell us what is epistemically justificatory. This answer assumes that truth is not itself epistemic. I take truth to be metaphysical, but have no further theory of truth to promote. This should not matter, because my expectation that epistemic justification admits of theorizing has a different motivation, one that does not depend on making a general case for the reducibility of the concept.

If epistemic justification were *not* analyzable, if the best one could do were to locate it within a circle of cognate epistemic concepts, then I would expect to find a much clearer conception of it in ordinary language. Our grip on the conceptually primitive is firm. Instead, I find the concept of epistemic justification elusive. This elusiveness is motivation enough to regard the concept as derivative, and a proper object of theorizing. An assumption about the nature of epistemic justification is necessary to identify the concept independently of theory, simply because the concept is *not* clear or univocal, in the way that, say, the concepts of knowledge, belief, or truth are presumed to be in treating them as coherent objects of analysis. Without specification of the *analyzandum*, any *analyzans* is in danger of being prescriptive and technical to the point of gratuity. Ordinary language is insufficient to constrain a theory of epistemic justification. A preliminary explication is needed to identify what we are to theorize about. This my assumption provides.

I say that I find the concept of epistemic justification elusive in a way that the concept of knowledge is not. Yet justification is normally regarded as a prerequisite for knowledge, as a necessary condition for knowing. My theory will leave open how justification relates to knowledge. Even if the normal view of their relationship is correct, this would not make knowledge derivative conceptually.[3] Knowledge is the more fundamental concept, and conceptions of epistemic justification arise as products of analysis, as creations of the philosophical art. Thus it is that, in general, intuition more readily decides whether one knows than whether a justification

---

one has none better served by true than by false beliefs. For one has no goal served by *any* beliefs. One can nevertheless value truth over falsity, in the way I describe. I maintain that, other things equal, one prefers not to be deceived. Of course this preference, even as I have qualified it, is ultimately a psychological matter. I cannot prove by philosophical argument that it applies to you. People who really do not care whether their beliefs are true, who have no preference for being right over being wrong, will not, on my theory, care about justification.

[3] Timothy Williamson (2000) gives arguments to show that A can be a necessary condition for B without being conceptually prior to B. I accept these arguments, although I would not have thought the point needed arguing. Williamson's specific concern is that knowledge be conceptually prior to belief. I do not go along with this, and will offer reasons for reluctance in Chapter 4. But what matters for my purposes is the comparison of knowledge with justification. My claim is that justification is not conceptually prior to knowledge, whether or not knowledge requires justification.

condition for knowing is satisfied. Suppose, for example, that one believes P because a person whom one has good reason to regard as a trustworthy authority, but who is in fact deceptive or misinformed, assures one of P. It happens, unbeknownst to one's source, that P is true. It seems clear to me that one does not know that P in this case, but unclear, pre-analytically, whether one's belief is justified; it is properly authorized but formed in a way that is in fact defective and likely to enmesh one in error.

There are also cases in which it is clear that one does know, but one's justification is questionable. A child knows where his mother hid the cookies; he can show you. But is it plausible to ascribe to the child a justified belief as to the location of the cookies on any basis other than the presupposition that knowledge requires justified belief? What of the dog who knows where he buried the bone? The paradigm indicator of knowledge is what one can demonstrate, in the sense of showing or doing. Of justification, it is what one can demonstrate in the sense of cogency of intellectual argumentation, a more elusive standard. In any case, these are distinct capacities.

Of course, connecting justification to the direction of credence toward truth is not the only way to pin the concept down. We should consider an alternative.

2.2 A Deontological Alternative

The concept of epistemic justification is elusive because the ordinary, pre-analytic conception of justification is not essentially, nor even primarily, epistemic. It has more to do with the morality of action than with belief or knowledge. If we analogize belief to action, treating believing as a kind of action, then ethics and action theory suggest an alternative philosophical provenance for an epistemic conception of justification. Rather than assume that (what I have identified as) the epistemic goal is advanced, one could follow this tradition and assume that one does no wrong in forming or holding one's belief. Doing no wrong in an epistemic sense could be promoted as an epistemic goal rival to believing truly (which, nevertheless, for me is *the* epistemic goal). The fidelity of this tradition to the ordinary conception of justification is, however, questionable.

Ordinarily, justification is at issue only where something is untoward; there is prima facie violation of a norm or expectation that constrains action. I go out to run an errand, leaving an infant unattended. The errand is of such importance that its neglect would be worse than any danger I can foresee to the child; I have alerted a reliable neighbor to check on the child; I have rigged an alarm should the child's protective enclosure be breached—these are justifications. Without one I do wrong. Running an errand leaving no one unattended (nor violating another trust or responsibility), I need no justification. It is not that my action is justified by the absence of responsibility. Justification is simply not at issue; there is nothing to answer for.

I see two possibilities. We can reject the analogy of belief to action: beliefs are epistemic commitments that require justification; actions generally do not. Or we can acknowledge a broad category of belief for which justification is simply inappropriate.[4] Much belief—perceptual beliefs, paradigmatically—are spontaneous, involuntary, unreflective. Only if they run into trouble or I put them to systematic intellectual use is there a need for justification. What does not work is to fault beliefs epistemically for want of justification while understanding justification on the model of moral accountability. Morally acceptable action is typically unjustified.

Therefore, justification as ordinarily conceived is inadequate to guide or check the philosophical ambition to theorize about justification in epistemology. An assumption is necessary. Demonstrating this need goes some way toward motivating the assumption I choose—that epistemic justification advances the epistemic goal of believing truly—if only by default. For the demonstration argues that the rival deontological tradition is not properly directed at justification at all. Deontology addresses blamelessness or permissibility, not warrant or sanction; perhaps "entitlement" is a neutral term. I am entitled to run my errand. I am entitled to do what I like if no moral constraint is violated. But nothing warrants my action. What would, that the errand needs running? Assume that it doesn't. There is nothing to justify. Admittedly, my purpose *explains* my behavior; I grant that, in general, actions have explanations. But explanations are not, in general, justificatory.

2.3   Difficulties with Deontology

What is worse, deontological concepts like blamelessness, compliance with norms, and the fulfillment of duty are ultimately incapable of supporting definitive judgments as to the justifiedness of belief. The problem is that there are different respects in which such deontological standards may apply to justification, and one may meet these standards in one respect while violating them in another.

To see this, let us ask whether the sincere, reflective conviction that one believes blamelessly, by any standard that it would be reasonable to impose, is inerrant. Suppose that one is blameless but doesn't think so, or to blame unrecognizedly. Could one's failure to recognize one's violation of constraints one takes to apply to one's investments of credence not be blameless? But then to identify justification with blamelessness becomes ambiguous.

A deontological assessment of justification proceeds from the subject's own perspective. Deficiencies in this perspective that the subject could not reasonably be expected to recognize or correct do not affect the subject's justification, deontologically understood. The subject's failure to take into account deficiencies inaccessible to him is blameless. Alvin Plantinga (1993, Chapter 2) argues, against Roderick Chisholm, that one can fulfill one's epistemic duty without achieving

---

[4] Instead, the inappropriateness of justification could be interpreted as justification easily won. On this view, justification is automatic until some epistemic norm is violated. I do not find this view at all intuitive, unless the norms in question require some kind of positive epistemic grounding, and will offer some criticism below.

positive epistemic status, because of unrecognizable defects in one's cognitive faculties. Although my argument differs from Plantinga's, his thesis may be thought to anticipate, partially, the point I am developing.

Plantinga's point is that one may have no epistemic responsibility for the failure of one's investments of credence to be truth-conducive. I concur, but my argument has nothing to do with possible defectiveness in one's cognitive faculties. One's perspective could be deficient not because of any cognitive defect, but simply because one's epistemic condition is not amenable to inerrant evaluation. Operating optimally, one's faculties may deliver beliefs whose epistemic credentials one blamelessly misjudges. One's epistemic condition, deontologically understood, is independent of one's own blameless evaluation of it. Plantinga assumes a (much higher degree of) privileged access to the satisfaction of one's own epistemic standards than my argument allows.[5] I contend that the subject's sincere belief that he is justified by all accessible measures may be mistaken, and the mistake itself blameless.[6] If this does happen, if, deontologically speaking, one justifiedly misjudges one's own justificatory status, then there is no coherent verdict as to one's justification.

I do not think it resolves this ambiguity to distinguish the justifiedness of first-order beliefs from the justifiedness of second-order beliefs about the justifiedness of first-order beliefs. It does not help to say, for example, that a first-order belief is unjustified, although the second-order belief that the first-order belief is justified is justified. For, one's justification for believing that a belief is justified would seem to justify the belief. It strains coherence to fault one as unjustified in believing what one is justified in believing justified.

There is further ambiguity at the level of judgments of justification. Whether or not a first-order belief is justified, the judgment that it is justified, no matter how well this judgment is justified by deontological standards, may be also be judged wrong by deontological standards. The problem is how, by deontological standards, a sincere, reflective, blameless assessment of justifiedness can be a misassessment. Deontology must assume that justified judgments of justifiedness are self-authenticating. But this assumption approximates rightly discredited claims of incorrigibility for first-person beliefs about the mental.

For consider: If I can be wrong about how things appear to me, then, presumably, I can change my mind about how I take myself to have been appeared to. Perhaps upon reflection I recognize something about my state of mind that I had failed to

---

[5] Plantinga is, to do him credit, uncomfortable with the assumption. See his Chapter 9, note 2.

[6] According to Laurence Bonjour's internalism (1985, Chapter 6), the accessibility of the coherence of one's belief-system is a precondition for justification. But then the truth of the beliefs this access provides—the metabeliefs that represent oneself as having a coherent system of beliefs—must just be assumed. Such beliefs are not assessable by the subject at all, let alone inerrantly. The accuracy of one's grasp of one's system of beliefs "must be taken for granted for coherentist justification to even begin" (p. 127). If Bonjour is right about this, then the consistent coherentist must admit that one can blamelessly mistake one's compliance with internalized standards of justification.

attend to. I thought that I was thinking of Venice when I contemplated that scene at the café by the water with the boats bobbing about, but now I realize that it must have been Bruges, for the waiter spoke Flemish. The example depends on an intentional state, one with reference to facts independent of myself. But I can also misassess my emotional states, for I can change my mind as to what properly constitutes them. I thought that I was afraid, when impending layoffs were rumored at work, but that wasn't really fear. Now that I've been taken hostage by Colombian rebels and held for a ransom my company has no intention of paying, I know what fear is.

All the more complicated, the more fraught with potential to err, is one's assessment of one's beliefs by the standards that one has reasonably, perhaps unavoidably, internalized. To hold oneself responsible beyond one's own measure of accountability is a confusion widely diagnosed. Evidently, I may believe myself guilty while also believing that this assessment is unwarranted.[7] I may require (and certainly would, were I a coherentist, require) my assignments of subjective probability to satisfy the axioms of probability, and believe blamelessly, but incorrectly, that they do. Violations of this requirement may be too difficult to detect by the standards of scrutiny that I hold it reasonable to apply. I may subscribe to and conscientiously implement a principle of total evidence, but my grasp of the concept of total evidence is understandably tenuous, and I am not to be faulted for failing to assign the weight that I myself believe is due to each of the data that I know of and take to be relevant. Further reflection upon what is already apparent changes my mind about what to believe. What belief, then, is justified? Was I precipitous, by my own standards, in investing credence? But I find, with further thought, that I was right to begin with. I am then revisited with doubt. Possession of justificatory status can no more await convergence upon some final, stable verdict than possession of truth-value can await the end of inquiry.

So, what is it to be? If people can mistake the features of their own mental imagery, surely they are not inerrant judges of the compliance of their beliefs with their own internalized standards. Because there is a difference between being blameless and blamelessly believing one is, there is a difference between believing justifiedly and believing in compliance with internalized standards.

Although the analogy of belief to action invites a deontological view of justification, my argument against viewing justification deontologically does not depend on the view arising in this way. One could just posit duties to constrain one's beliefs, and count as justified beliefs that comply with the constraints. This view is deontological whether or not beliefs are like actions, for it makes justification a matter of fulfilling one's duties. Dissociated from the morality of action, I do not see what the motivation for the position would be, but never mind; maybe you do. The point still

---

[7] I think it sufficient to appeal to common experience here, but I also note that the condition I describe has a compelling realism in great literature. An example is the character of Gwendolen Grandcourt in George Eliot's *Daniel Deronda*. Gwendolen blames herself for her husband's death, because she blames herself for having wished him dead. At the same time, she believes this self-condemnation to be excessively harsh.

holds that one may justifiedly misjudge one's justificatory status; justification has not been internalized.

For even if one must be able to do whatever is one's duty, even if one cannot have duties that one is unable to perform, it does not follow that one must be able to determine whether one's duty has been done nor what one's duty is. One may incorrectly assess one's performance without this mistake constituting or reflecting a failure to act dutifully. Although obligation implies ability, blamelessly believing one is obligated does not imply ability. It is not plausible to impose, whenever there is a duty to do x, a further duty to tell, correctly no less, that one has done x. The physician may have a duty to do all he can to save his patient, and doing all he can is certainly within his abilities. But need he be able to determine that what he has done for the patient *is* all that he can do, that no further measure would avail? It is hard to see, in general, how one could know this. It is easy to see how the physician could miss something he could have done, or expect more of himself than is possible. One may blamelessly require of oneself what one is unable, and hence not required, to do. Moreover, if there were a duty to determine that one had done one's duty, then there would be a further duty to determine that one has determined this, and so forth. But it is not plausible to maintain that one's duties ordinarily comprehend such a regression.

Understanding justification as the fulfillment of duty does not internalize justification unless the specific duties imposed happen to be ones, if any there be, whose fulfillment is automatically within the ken of the agent. Compliance with duties that constrain one's assessment of one's fulfillment of one's epistemic duties does not ensure the accuracy of this assessment. For, there cannot be a duty to assess one's compliance accurately. In general, the agent can justifiedly misjudge his justificatory standing, deontologically understood. The agent who does his duty may justifiedly judge his performance wanting. And the agent who fails to do his duty may justifiedly judge himself compliant. Justificatory status proves irremediably ambiguous on a deontological view.

2.4 Justification and Argument

There is the complication that being justified, having justification, and justifying all differ. One may have a justification without realizing that one does, or without making any use of it to justify anything. One may have a justification for believing a proposition that one does *not* believe, or for a belief that one holds for altogether different, possibly nonjustificatory reasons. One might take advantage of these distinctions to single out a restricted form of justification that can be understood deontologically. I say not that this is impossible, but that its verdicts as to justifiedness would be stipulative and incongruous with results equal in deontological motivation.

Moreover, at least one form of justification strikes me as strongly resistant to deontological interpretation on straightforward intuitive grounds. The concept of justifying a belief as an action one takes—the giving of reasons, the citing of evidence, the refutation of opposition—is by nature inhospitable to deontology. Justifying in this sense is arguing. And *what* one argues, in justifying one's belief, is *not* that the belief is held blamelessly, but that it is *true*. The point of justification in this sense is not about oneself or one's entitlements at all, not about one's holding of the belief but about what is believed.

Naturally, the deontologist thinks that believing in accord with the internalized standards that impose epistemic duties is truth-conducive. He may counter that the point of satisfying deontological constraints on belief is to get truth. But what is noteworthy is that in justifying one's belief one does *not* attempt to show that deontological constraints have been satisfied. One argues that one's belief conforms to the world, not to one's standards.

These considerations seem to me to favor a conception of epistemic justification that connects directly with truth. We may call such a conception "externalist", in that failure to connect with truth does not, in itself, imply any violation of internalized standards as such; one may not have internalized the right standards. Nevertheless, so long as one *takes* oneself to have internalized truth-conducive standards, one is in a position to defend one's beliefs, to construct what, for all one, perhaps anyone, might be able to tell, are justifications of them. The structure of one's defense places the satisfaction of one's internalized standards in the premise position, and places truth in the conclusion. From the fulfillment of epistemic duty, one argues for truth. This generates a notion of justification that is internalist in that its implementation in argument does not depend on a connection to truth having actually being effected. The premise need not be true.

This notion may be compared to the notion of explanation credited to scientific theories independently of their espousal. A theory is said to explain in the sense that it offers or proposes *an* explanation; it would *really* explain if it offered the correct explanation. Rival theories offer competing explanations, so each offers an explanation and to that extent explains. But at most one of the theories really explains.[8] Compliance with standards one takes to be truth-conducive offers a justification that really justifies if the standards are right. This conception of justification is clearly derivative; fundamentally, epistemic justification is externalist.

I began with the assumption that epistemic justification is justification that advances the epistemic goal of believing truths without believing falsehoods. Some assumption as to the nature of epistemic justification is needed to explicate the concept preparatory to theorizing about it. For ordinary thought entangles the concept with deontological notions that are not essentially epistemic. I hope now to have motivated my assumption and the externalism implicit in it. Let us then confront some consequences.

---

[8] This is not to say that correct explanation is unique. Explanation has contextual and pragmatic aspects, so that different explanations of the same phenomenon can be correct for different purposes. But among rival, incompatible theories offering competing explanations, at most one explains correctly.

2.5 Connecting Justification to Truth

An immediate point is that the property of advancing the epistemic goal—*truth conduciveness*, let us call it—does not, speaking strictly, attach directly to beliefs, as such. And yet beliefs are the objects of epistemic justification.[9] A belief is true or false; what would it mean for a belief to be truth-conducive?

It could mean that the act of believing it (if we may analogize believing to acting; the *holding* of it, to be careful) leads one inferentially, causally, to further beliefs that are or tend to be true. But surely a belief can be justified without inducing further beliefs at all. And an unjustified belief, a false one, for good measure, could easily be truth-conducive in this inferential sense. The unjustified and false belief that a reference work—let us make it the *Encyclopedia Britannica*, for future reference—is inerrant could lead one to believe lots of truths. (Perhaps someone has high standards and would not use the encyclopedia unless he believed it inerrant.)

Surely this possibility is congenial. Justification is not truth; the former concept (in application to beliefs) is epistemic, the latter metaphysical. Justification is achieved, paradigmatically, in science, and science gets things wrong. We do not want a conception of justification that requires, let alone collapses into, truth. Nor is justification truth-conduciveness. We should expect there to be cases like the encyclopedia, because justification and truth-conduciveness can vary inversely.

Earl Conee (1992) has an example in which one has no reason whatever to believe that P is true, but has good reason to believe that he will be in a position to determine whether or not P is true if and only if he first believes P. Conee decides that believing P under these conditions is justified, but that P itself is unjustified. I prefer to say that believing P is unjustified despite its truth-conduciveness, because justification does not attach to the process by which the epistemic goal is advanced but rather to the beliefs that advance this goal; that is, to the beliefs in which its advancement consists (the beliefs one gets in advancing it). The truth-conduciveness of believing, whatever the truth-value of one's belief, does not justify the belief; justification is to be bestowed, if at all, on the beliefs that truth-conducive believing induces.

Consider a scenario in which P is true and one has strong indication of P's truth, but the effect of believing P is deleterious with respect to the epistemic goal. Maybe believing P fuses synapses in the brain, or even (Conee, 1992) triggers an explosion that puts one out of the epistemic enterprise permanently. Here I think it important that, nevertheless, believing P *does* advance the epistemic goal; it just does not do so *on balance*. The net effect is regressive. But we cannot require for justification that the goal be advanced in the long run, because the impossibility of this would not pre-

---

[9] So I am assuming. There are other possibilities. Justification could attach to *revisions* of belief, or to adjustments in degrees of confidence, obviating belief altogether. I shall say something about such alternatives in the course of developing my own theory of the justification of beliefs (see especially Chapter 5).

empt justification. After all, reason itself could be disadvantageous in evolutionary terms. What if the capacity to weigh evidence depends (nomically) on mental traits that are ultimately self-destructive? Still, evidence can justify beliefs by directing credence toward truth.

But the more important point about this kind of scenario is that it sustains a bifurcation between the objects of justification and truth-conduciveness. Where believing P generates further beliefs, how believing P fares as to truth-conduciveness does not decide the justificatory status of believing P but only that of the further beliefs generated. Whether believing P is justified depends not on its effects, but on its causes.[10] Accordingly, I propose to regard truth-conduciveness as a property not of beliefs, but of methods or processes of forming beliefs, induction by antecedent belief being one possible process. And justification is not to be identified with truth-conduciveness; rather, justification attaches to beliefs truth-conducively produced. And truth-conduciveness is not to be identified with the production of truths; it must be possible for false beliefs to be justified. Believing truly automatically advances the epistemic goal somewhat (supposing, again, the interest or importance of what is believed). But truth cannot be necessary to justification.[11]

This seems to me a claim that it is appropriate to assume, not to defend. I assume that the concept of epistemic justification applies to science. It cannot incorporate truth, because science at its best can be wrong. Your agenda might be different. What you might want from justification is a property that turns belief into knowledge, whereas what I want has nothing directly to do with knowledge. I am after a property that beliefs have in virtue of advancing the epistemic goal. As I understand this goal, it does not require knowing. We make epistemic progress by being right without being wrong, even if we do not have knowledge.[12]

Of course, one (e.g., Conee, 1992) could start with knowledge as the epistemic goal rather than truth, perhaps because being right by luck should not count as epistemic success. I do not have to refute this priority, because the theory of justification that it generates need not conflict with mine. However, I do have reason to think

---

[10] Marian David (2001) argues that justification cannot depend on effects, but infers that it must then be synchronic. Against Foley (1987, 1993), he complains that a synchronic account identifies justification with truth. But justification dependent on the etiology of belief is not synchronic. The goal *now* to believe all and only what is true is not the goal *now* to believe all and only propositions with a certain causal history. This would be incoherent, since there is such a history only for what one already believes.

[11] Is truth sufficient for justification? Is any truth trivially justified in that the holding of it advances the general epistemic goal *somewhat*, and no greater, net advance can be required? Well, from a (thoroughly) externalist perspective, that a belief is true *is* justificatory, if other things are equal. But I do not think that the kind of justification it represents attaches to the believer. Further consideration of this issue must await the development of a theory. It is an issue that I think it takes a theory to resolve.

[12] Michael DePaul (2004), following Conee, seeks an intrinsic value for justification independent of its instrumental utility in directing our beliefs toward truth. DePaul does not consider the possibility that the independent value of justification is to serve the goal of monitoring our progress with respect to the goal of truth.

that the priority should go the other way. I suggest that being right has intrinsic value, and that the reason we value knowledge *in addition* to being right is that we want not just to advance the epistemic goal but also to ascertain this is what we are doing. Because we would want to be right even if we could not tell that we were, truth as the epistemic goal stands independent of the further desideratum of knowing.

But even if your agenda is knowledge rather than truth, I do not think you can build truth into justification. Knowledge is partly metaphysical, requiring the co-operation of the world, and cannot be explicated entirely in epistemic terms. What you *can* do, with knowledge your target, is limit consideration to justified truths. But then, as observed in Chapter 1, you are *bypassing* the concept of justification as such. You do not need a general theory of justification. As known propositions are true, it suffices for you to say what it takes for truths to be justified.

This is a further reason why justification is a more elusive target than knowledge, despite its presumptive status as a condition for knowledge. In the tradition that asks what *additional* conditions are needed to close the gap to knowledge, justification tends to get bypassed, relegated to unexamined intuition. The analytic problem for this tradition arises once justification is granted. Thus, little analytic attention to justification is to be found in the enormous literature generated by the recognition that a justified belief could be true for reasons unconnected to its justification, and so not known.

This recognition is credited to Edmund Gettier (1963). A different response to Gettier, represented, for example, by Fred Dretske (1971), is to strengthen the justification condition for knowledge, to contend, in effect, that in Gettier's cases the subject is *not* justified after all, or is justified only *weakly*. Any viability conceded to this response is further indication of the intuitive instability of justification. Latitude to adjust the justification condition reinforces my contention that epistemic justification is a more elusive analytic target than knowledge, requiring of special pre-theoretic explication.

2.6 Falsity-aversion

Another point, perhaps not immediate, is that it must be possible, in advancing the epistemic goal, to avoid falsity altogether. Truth-conduciveness, as I understand it, is falsity aversive. What if there is no systematic way to form beliefs that gives one even a *chance* to block falsehoods altogether; the cost of getting lots of truth is that *some* falsity, perhaps very little and seemingly minor by comparison, is *inevitable*? Then the epistemic goal, as I have identified it, is unachievable. No method is truth-conducive; according to my assumption, there is no epistemic justification.

Does this sound harsh? Perhaps a very little bit of error seems a fair exchange for lots and lots of truth, to the point that we should count such an outcome, if it is the very best that we could ever do, as the achievement of our epistemic goal. If this is your view, we disagree.[13] Of course, I cannot prove you wrong. (That's why I need an *assumption*.) If you are willing not merely to *risk* error, to endure the *likelihood* of error, in your quest for truth, but are acquiescent in its *ineliminability*; if you think the very *goal* of epistemology tolerates error without prospect of redemption; well, then I beg to differ. I am inclined to suppose, however, that the real disagreement is more tractable. You think that there are or may very well be no methods or standards of investing credence that reduce error to a risk, and I don't. I think that justification as I conceive it is achievable; you don't. If this is the difference between us, perhaps there is hope for you in the theory of justification I have to offer.

I must, however, acknowledge, as an implication of my assumption, that certain appealing modes or standards of belief-formation cannot be justificatory. High probability, beyond any threshold (<1, of course), is an example; the members of an inconsistent set of propositions can have individual probabilities as high as one likes. The high probability of a proposition cannot, *in itself*, justify believing the proposition, for then false justified beliefs would be inevitable.

Well, perhaps not quite inevitable, for one might not invest credence where one is justified in doing so. One might fortuitously fail to believe falsehoods by *happening* to arrest credence just when one gets to them. But in this case, one has ceased to pursue the epistemic goal. My assumption is that it must be possible to pursue the goal without coming to believe falsehoods, and I will show that this assumption disqualifies high probability as a sufficient condition for justification. In particular, it will turn out that with high probability as a standard, not only are falsehoods unavoidably justified, but so are overt contradictions (i.e., of the form $P\&\sim P$). This result violates a condition of adequacy adopted in Chapter 1. Accordingly, high probability is not justificatory on the theory of justification I will propose. At most, high probability justifies believing that a proposition is highly probable, which must be distinguished from belief *simplicitor* (from believing that the proposition is true).

As another example, consider Tarski's truth condition for sentences S: "S" is true if and only if S. This is sentence schema, not a sentence. As it does not make sense to speak of believing a schema, I do not have to *dis*believe it, or take issue with it as an explication of truth. As it cannot be a belief, it is ineligible for justification. But its instances are eligible, and they cannot all be justified. It cannot be a truth-conducive method of belief-formation to believe all instances of the schema, or to take instantiation of it as justificatory. For some of its instances are contradictory, and some not contradictory are collectively inconsistent.[14]

---

[13] Richard Foley (1987, Chapter 2, Section, 2.3) diagnoses the avoidance of false belief in epistemological theories as an "obsession" that unwittingly encourages skepticism. It seems to me that the avoidance of skepticism in epistemological theories is an obsession that unwittingly encourages acquiescence in error.

[14] For S_1: "S_1" is false, the instance obtained by substituting S_1 for the first occurrence of S and its referent for the second occurrence of S is contradictory. For S_1: "S_2" is true, S_2: "S_1" is false, the instances obtained by substituting S_1 and S_2 for the first occurrence of S and their referents for the second occurrence of S are inconsistent.

2.7 Propositional Justification

I have limited justification to beliefs. Propositions are believed, but justification attaches to propositions only in so far as they are believed. It might make sense to conceive of justification more abstractly: a proposition is justified if a truth-conducive way of coming to believe it is (potentially?) available, and were it to be believed in this way the resulting belief would be justified. One might think of justification as a kind of ground or warrant possessed by or available for a proposition independently of whether any prospective believer recognizes it or acts on it to form a belief.

But such a conception is in danger of subsuming truth under justification. If truths have *truth-makers*, conditions that cause their truth, then presumably there exists a truth-conducive way, in principle, of coming to believe any truth: recognition of whatever makes it true. Then every truth is in principle justifiedly believable. So every true proposition is justified. Perhaps some truth-makers are in principle unrecognizable, but I would not hold the independence of truth from justification hostage to this metaphysical possibility.

I grant that there is precedent in ordinary language for extending justification to unbelieved propositions. We can speak of intellectual stances being justified irrespective of being advocated. The jury cannot reach a verdict, but the prosecution contends that a guilty verdict is justified. I propose to interpret such talk hypothetically: if the proposition were believed, this belief would be justified. This interpretation cannot consistently be carried through at the second order. For example, the evidential basis for the proposition that a certain idea has never occurred to me could be as strong as one likes. But we cannot say that if I came to believe this proposition my belief would be justified, for my belief must then be false. Second-order limitations will concern me later (primarily in Chapters 10 and 11). I think it suffices here to note that ordinary language exerts little pressure at that level.

There is a further point. For every belief there is a believer, perhaps more than one. Suppose two subjects believe the same proposition, but form or sustain the belief differently; one truth-conducively, one not. The question of whether the belief *itself* is justified becomes ambiguous. We must be prepared to distinguish one's belief from the other's, although, as there is only one believed proposition, in a natural sense they have the same belief. This complication seems to me allowable as an instance of the familiar distinction between qualitative and numerical identity. Beliefs identical in semantic content can differ as to justification.

I hope now that we have an idea of epistemic justification clear enough and plausible enough to support theorizing.

Chapter 3
Reliability

3.1 The Core Notion

I propose to explicate truth-conduciveness in terms of reliability. In this chapter I shall be concerned both with the reliability of processes or methods by which beliefs are formed, and with the reliability of the formation of beliefs. These are distinct notions. Each incorporates a condition unique to the theory of justification I shall propose: The reliability of methods introduces a condition of normalcy, and the reliability of belief-formation introduces a condition of intentionality. Much of this chapter is devoted to explaining these conditions and applying them to problems raised by the notion of reliability.

To form or sustain beliefs reliably is to form or sustain them in a way that can be trusted not to deliver falsehoods. This is the core notion of reliability. Of course, the strongest basis for this trust would require that the resultant beliefs be unqualifiedly true; it would require an inerrant method. What we rely on a method for is to get things right, and it deserves our trust if it does this. But if the truth-conduciveness of method is to explicate justification, we cannot require so strong a basis for trust. For false beliefs can be justified.

I see two ways to develop the core notion of reliability as trustworthiness of method so that it provides for false, justified beliefs. One way emphasizes the acquisition of truth. Our method yields lots of truth and little falsity; in investing credence reliably, the preponderance of truth over falsity is high. This is Goldman's way (1979), discussed in Chapter 1, and it has become the standard, probabilistic interpretation of reliability.[1] The other way emphasizes avoidance of error. While not infallible, reliably formed beliefs can be counted on rarely if ever to be false,

---

[1] As I have noted, Goldman also applies the term 'reliability' to subjunctive conditions for knowledge (1976). Ernest Sosa and William Alston (1995) are further proponents of the probabilistic interpretation. Sosa's position(s) will be considered in Chapter 9. The probabilistic interpretation is sufficiently prevalent that Carl Ginet could begin his (1985) *critique* of reliabilism by just assuming that "the reliability of a belief-producing process is a matter of how likely it is that the process will produce beliefs that are true." Earl Conee and Richard Feldman (1998) make the same assumption in their critique of reliabilism.

J. Leplin, *A Theory of Epistemic Justification*, Philosophical Studies Series 112, DOI 10.1007/978-1-4020-9567-2_3, © Springer Science+Business Media B.V. 2009

though perhaps at the cost of being few. As I have assumed it to be part of the epistemic goal to avoid falsity altogether, I shall take my lead from the latter conception of reliability.

Reliability in the sense of high relative frequency of truth to falsity as output is not justificatory, because it guarantees that some resultant beliefs will be false. Assuming that frequency is measured over beliefs actually formed, and does not count counterfactual applications of a method in which a false belief would have been formed, a frequency below 1 makes false belief unavoidable. This problem is independent of the lottery problem. With probability the standard of justification, the lottery problem is introduced by making reasoning a justificatory process. One reasons from probabilistically justified beliefs, each to the effect that an individual lottery ticket will lose, to a conclusion whose negation is independently justified, and so to a contradiction. But *any* process that is justificatory in virtue of issuing a high frequency, short of 1, of true to false beliefs guarantees false justified belief. If all the outputs of a method reliable in the frequentist or probabilistic sense are justified, then some false beliefs are justified. As I understand the epistemic goal, this result is unacceptable. Reliability must *provide* for false justified beliefs *without* guaranteeing them. If a method is justificatory, it must at least be *possible* that all beliefs the method yields are true. While it is consistent with the justificatoriness of a method that it yield false beliefs, it is not consistent with the justificatoriness of a method that it render false beliefs inevitable.

Strictly speaking, this restriction does not immediately preclude a probabilistic interpretation of how reliability is justificatory. One could count a method reliable if the probability that the beliefs it issues are true is high, but give probability a different, nonfrequentist interpretation.[2] Then, strictly, it will be possible to use the method without generating false beliefs. But this approach proves unacceptable, for a number of reasons. Some reasons will emerge in subsequent chapters (especially Chapter 6). The immediate objection is that however improbable this approach renders the generation of any particular false belief, the avoidance of falsity altogether becomes an overwhelmingly improbable accident. While infallibility cannot be expected of it, a justificatory method should be error-resistant, not just in the sense that errors are few but in the sense that they are, taken individually, *resisted*. The method must in some way be sensitive to error. As I understand the epistemic goal, a method that advances this goal must not leave the avoidance of error entirely to chance.

For this reason also, I do not count false beliefs formed in counterfactual applications of a method in determining frequency. That a false belief which would have been formed happened not to be does not qualify a method as error-resistant in the sense required for justification. We want truth for the beliefs that a justificatory method would deliver, as well as for those that it does deliver.

So how can a method be "counted on" rarely if ever to yield false beliefs, without precluding them altogether? The key intuition is that methods have natural or

---

[2] For example, Goldman sometimes (e.g., 1979) speaks of propensity as well as frequency.

intended ranges of application. No method of doing anything can be expected to work successfully under all possible conditions. The classic example, from David Armstrong (1973), is the use of a thermometer to determine temperature. Thermometers have ranges of sensitivity, depending upon their design and purpose. No thermometer will read accurately at all physically realizable temperatures. Mine is an ordinary outdoor thermometer with a range of $-40°C$ to $+50°C$. At lower or higher temperatures it is not to be trusted, but such extremes do not normally occur. Of course, I cannot use my thermometer to verify that the temperature is not extreme, for if it were my thermometer would not so indicate. Rather, in using my thermometer I *presume* that the prevailing temperature is within the normal range. Perhaps, additionally, I reason that there would be independent indications of abnormalcy, but these indications pertain to a *different* method. With respect to the determination of temperature by the reading of my thermometer, the absence of extremes is presupposed. This presupposition could be mistaken. I count on the thermometer rarely if ever to yield false beliefs about the temperature, in that I count on the presuppositions of my use of the thermometer rarely if ever to be mistaken.

I will say that a process or method of belief-formation is *reliable* if it would not produce or sustain false beliefs under normal conditions. If a belief is produced or sustained by a reliable process under normal conditions, then were this belief to have been false, the process would not, under those conditions, have produced or sustained it. This is a subjunctive form of reliabilism of the sort Nozick proposed for knowledge, discussed in Chapter 1.

The underlying subjunctive condition, absent the restriction to normalcy, has been labeled "sensitivity" (e.g. by Sosa, 1999) and distinguished from its contrapositive, "safety": a reliable process would produce or sustain a belief only if the belief were true. I will consider the safety condition and how it relates to sensitivity in Chapter 8. For now, I note that the sensitivity of a method has not been intended to exclude all possible worlds in which the method forms a false belief. The worlds in which false beliefs are formed will have to have something wrong with them, if they are not to interfere with the method's sensitivity. One might identify this failing, whatever it is, with abnormality, and contend that my restriction to normalcy is already implicit in the sensitivity condition as ordinarily understood. I suppose this is all right, as far as it goes. But I wish to make the restriction explicit and to explain how normalcy should be understood. As I shall develop the notion of normalcy, it has little to do with notions like relevance, proximity, similarity, or salience, which have been used to identify the accessible worlds.

The reliability of a process, understood subjunctively, does not require that truths result in high proportion or with high probability. Nozick's version does carry this consequence, but mine does not. The difference results from further conditions he adds, and from the absence from his version of my restriction to normalcy. As Nozick seeks conditions for knowledge, the issue of the desirability of this consequence does not arise for him. He needs a truth condition anyway, so he can let the probability of getting truth be 1. My version of subjunctive reliabilism will differ substantially from Nozick's, and will be developed, in the first instance, for justification rather than for knowledge.

The immediate novelty of my version is its definitional reliance on the notion of normalcy of conditions. In calling a belief "reliable", I mean that it issues from a reliable process. Then a reliable belief may be false if conditions are abnormal. Note that normalcy is *not* a condition for the reliability of a method. Rather, it is a condition for a reliable method to be inerrant. It may seem natural to regard consulting my thermometer as an unreliable method at $> +50°C$, for it is not trustworthy at such temperatures. The best I can do to accommodate this intuition is to pronounce one's reliance upon it mistaken. One may, if one wishes, judge the thermometer *itself* unreliable at $> +50°C$. But I find it convenient to abbreviate the reliability of the method of consulting an instrument by the reliability of the instrument.

Note further that reliability, as subjunctively defined, carries counterfactual consequences. To *establish* that a method is reliable, it is insufficient to verify its actual results within its range of normalcy. It is necessary, somehow, to determine what results it would give under hypothetical circumstances that do not occur. Is my thermometer reliable; that is, is consulting my thermometer and believing what it reads a reliable method of forming beliefs about the temperature? It is if my thermometer is working properly (not stuck, for example, even on the right temperature). Then, under the hypothetical circumstance of a different temperature the thermometer reads that temperature. A law of nature proportioning the height of the mercury to the temperature guarantees this; natural laws sustain counterfactuals. In other cases it may not be so clear how to assess reliability. Of course, a method can be reliable without its reliability being established or (even) establishable.

Notice that the subjunctive mood of the definition obviates the possibility that a method is trivially reliable in virtue of yielding *no* results. A method is a method for doing something. If not only are no results yielded but also none would, hypothetically, be yielded were it (systematically) used, then I do not see that it makes sense to speak of a *method* at all.

3.2 Explicating Normalcy

Naturally, the notion of normalcy requires explication. There are different things that it could, without violence to intuition, mean, with different consequences for reliability. Normalcy is relatively straight-forward in the case of an artifact designed for a specific purpose: normal conditions are those it was designed for, and laws of nature determine what designs will work. But lots of methods of belief-formation that we do and must trust are not artifactual, and those that are depend on others that are not. The thermometer depends on perception. Perceptual faculties are the product of evolution, not design. It is not part of my method of forming a belief about the temperature to form a belief about the thermometer's reading, but unless perception were a reliable way to form beliefs about its readings consulting it would not be a reliable way to form beliefs about the temperature. What does normalcy mean, in general?

One possibility is a frequency interpretation: normal conditions are those usually in effect, or usually in effect on occasions of and in the environment of the method's use. But this threatens to reintroduce a frequency interpretation of reliability. Reliable methods usually give the right results because they are usually used under conditions that are (by definition) normal in which they do not (by definition) err. Perhaps what happens under unusual conditions is not that we get a false belief but that we get no belief. The thermometer does not read incorrectly; it breaks and the mercury leaks out. Then false belief is not inevitable under a frequency interpretation of normalcy. This is not reassuring. What if the temperature is 55°C, not enough to break the thermometer; it stays at 50°C, its highest reading, and resumes accurate performance when the temperature falls? As I reject the frequency interpretation of reliability, I also reject the frequency interpretation of normalcy.

A possibility I like better is that normal conditions are those under which the concepts ingredient in beliefs the method produces are acquired. These conditions are likely to be the same as those under which the method is used, but are not necessarily the same and cannot be identified with conditions of use. One might grow up in Hawaii, become a government oceanographer, and get assigned to Antarctica. By extrapolation, changes could be so great that normal conditions never obtain (any longer). Someone who, ignorant (somehow) of such change, continues to use the method could err systematically, despite the reliability of the method. The presumption of the method's use, then, is that there not have been wholesale, unrecognized changes of environment that vitiate one's results. That would be a skeptical scenario that one presumes not to be in play when one decides almost anything.

The difficulty with this proposal is that I do not know how to keep concepts intact across hypothetical changes in the conditions under which they are formed. To define normal conditions as those under which concepts *happen* to have been formed, and which might have been different without effect on *what* concepts are formed, may assume too internalistic a picture of meaning. I would not assume that denizens of Twinearth have our concept of water.[3] I myself am dubious of what residents of Minnesota, let alone Antarctica, mean by a "nice day".

A more promising account is that conditions normal for a method are conditions typical or characteristic of occasions and environments in which the method is usable or applicable, whether or not it is in fact then or there used or applied. My thermometer is not (for the most part) usable under extreme conditions. (Ordinary) perception fails in the dark. It might happen that on all actual occasions (and in all actual environments, hereafter implicit) of a method's use, the conditions are *atypical* of occasions on which it is usable. Then, despite the method's

---

[3] What is called 'water' on Twinearth is superficially like water, and would normally be taken for water if introduced on Earth, but is structurally different. So the concept 'water' on Twinearth does not refer to water, and that (it may be argued) makes it a different concept. The example is due to Hilary Putnam (1981), a pioneer of semantic externalism.

reliability, the preponderance of the beliefs it yields could be false. Truth could be infrequent among beliefs that a reliable method delivers. Only on the supposition that reliable methods are rarely if ever used under abnormal conditions are reliable beliefs rarely if ever false. But they *need* never be false, for conditions need never be abnormal.

This result should obviate the worry that the notion of what it is for conditions to be typical or characteristic of occasions of a method's usability adverts, once again, to frequency. A frequency notion at this point does not reintroduce a frequency interpretation of reliability. A method inerrant under conditions frequent when it is used, with frequency <1, will, infrequently, be used when errant. It yields falsehoods infrequently but unavoidably, and so is not truth-conducive in my sense. But a method inerrant under conditions frequent when the method is usable need never be used when errant, and so need never yield a falsehood. Inerrancy under conditions frequent when a method is usable constitutes a form of reliability suited to a falsity-aversive epistemic goal.

Moreover, being typical or characteristic is not essentially a frequentist notion. Conditions that typify or characterize occasions of a method's use speak not just to whatever happens to obtain when the method is used, but more specifically to what it is about such occasions that enable the method then to apply. Typifying or characterizing conditions identify features of the occasions in virtue of which the method is usable. To be typical or characteristic is a vague notion (and not quite univocal in ordinary language), but I shall sharpen it up through consideration of examples.

Suppose that I have acquired concepts of emotions through common social interactions, but I have formed few beliefs as to actual incidences of certain emotions. Exposed to a company of actors, I take them for regular people (not acting, anyway); their (apparent) emotions are, unbeknownst to me, feigned. I form lots of false beliefs about their emotional states, my method being to judge by behavior. The method is reliable. The existence of actors does not make it unreliable; rather, the conditions are abnormal. It is atypical of human interaction for behavior *not* to be indicative of emotional states. Otherwise, having only behavior to go on, we would not have reliable beliefs about one another's emotional states; society would be nothing like we know it. Notice that behavior's indicativeness of emotion is not simply a matter of frequent concomitance. We assess emotions by behavior not because of an accidental regularity but because behavior is expressive of emotion. It is *characteristic* of being in an emotional state to behave in a certain way.

What if actors take over (some becoming politicians)? What if feigning emotions becomes commonplace and transparency the exception? Notice that this would be, not simply a shift of patterns of concomitance, but a misrepresentation of emotions by suppression of their characteristic expression. If this happens, then eventually, I suppose, my method would become unreliable, but not right away. Immediately, affectation, though predominant, is atypical and uncharacteristic. And to imagine it becoming characteristic is to imagine some transformation in emotional states themselves.

A classic example, from Alvin Goldman (1976),[4] is about barn facades. Judging by appearance can be a reliable way to identify barns, although it yields false beliefs in a region replete with fake barns (mere facades). My point, in terms of this example, is that it might only be in this region that one comes to identify barns visually, so that the great preponderance of one's beliefs as to the presence of barns are false. The method is nevertheless reliable, as the region is abnormal. Notice, again, that its abnormality is not *simply* a matter of frequency. Fakes do not merely *happen* to be the exception; at least in the long run they are conceptually required to be exceptional.[5]

One might nevertheless protest that frequency has a lot to do with it. The abnormality of fake barns depends on how prevalent are facades in relation to real barns, and how widespread is the mixture. Is it not arbitrary at what point deceptive conditions become normal? But if the division between normal and abnormal conditions is unclear, so is reliability.

I do not think that these questions are the right response. The arbitrariness is not in the notion of reliability, but in the example. In taking what looks like a barn to be one, it is presupposed that there is no local incentive to trickery as to the existence of barns, that this is not the set of a Hollywood western. The recipe for manufacturing false reliable beliefs is to hypothesize credible violations of the natural presuppositions of the use of a method. Placed in an alien environment, the subject misjudges things systematically. This is a recognized dramatic form. Both the abnormalcy of the conditions and the subject's insensitivity to their abnormalcy are understandable. If the recipe is not followed the results are strange, intuitively less clear. A scenario that does not even attempt to make plausible the conditions that sabotage one's investment of credence, that places fake barns where there is no reason, concealed from the subject, for them to be, simply *posits* abnormalcy. Rather than fault reliability, we should just register dissatisfaction at having to accept abnormalcy as a brute posit. We should insist that the recipe be followed. The example should then be rich enough to decide, nonarbitrarily, the normality of deception.

We have now identified two related criteria of normalcy. To avoid ambiguity, I shall propose an explication that makes one criterion official and one informal. Officially, normal conditions are those characteristic of situations in which a method is usable to form beliefs. Characteristic conditions frequently are satisfied when the method is usable, because they are conditions that contribute in some way to the method's utility, or usefulness. And unless a method is useful, as well as usable, unless we want the results it delivers, we are disinclined, other things being equal, to use it. But normalcy does not require utility.

---

[4] Goldman credits Carl Ginet.

[5] I realize that this is quick. Its credibility depends heavily on the vague qualification that this happens in the long run. Here I can only add that cases in which actual x's are rare in relation to things that can be (mis)taken for x's are not generally cases in which the later are fakes. Reproductions of art works are not fake, for example.

The motivating idea is that no method is applicable universally, independent of prevailing conditions, and the motivating examples are perceptual faculties like sight and artifactual instruments like thermometers. These carry unproblematically circumscribable ranges of application, outside of which they are unusable. For more sophisticated methods reliant on judgment and experience, like the weighing of evidence or testimony, limitations of applicability are less definitive. For these methods, the ready criterion of normalcy is what is presupposed by their use. Informally, normal conditions are the conditions the subject presupposes in using the method.

In judging color or shape by sight, we do not presuppose that the objects of vision are illuminated; without illumination the method simply does not operate. Illumination is a precondition rather than a presupposition, because no judgment about illumination is made (even implicitly). In trusting expert testimony we do presuppose ingenuousness, precisely because it is possible to trust the disingenuous. Of course, in judging by testimony, assessing honesty and expertise is part of the method. Simply to assume these traits in a testifier is gullibility. But in believing on the basis of testimony we assume that we have not been manipulated by the unscrupulous or deceived by the unqualified. As a precondition, illumination meets the official criterion of normalcy for belief by sight. As a presupposition, ingenuousness meets the informal criterion of normalcy for belief by testimony.

The difference in these cases is, however, one difference of emphasis. Vision has presuppositions, like the absence of trick mirrors, and testimony has preconditions; it is unavailable to the isolated. Presupposition carries the greater burden of determining normalcy the greater and more varied are the dangers that a method is used under conditions that subvert it. The two criteria connect via the assumption that systematic and routine violations of a method's presuppositions ultimately render it unusable, much in the way that induction breaks down in a chaotic universe. Granting this assumption, the standard of normalcy is univocal.

3.3 Problems with Normalcy

It might be objected that a method does not become unusable under conditions that violate its presuppositions; if it did no false beliefs could be formed by a reliable method. Why, then, cannot violations of presuppositions be characteristic of situations in which a method is usable? Rather than unusable, the method would be rendered unreliable. But then abnormal conditions become conditions under which the method is unreliable. In effect, a notion of reliability is implicit in the analysis of normalcy, which, in turn, is being used to explicate reliability.

I reply that normalcy is analyzed without circularity as conditions consistent with the preconditions for and presuppositions of a method's use. These are the conditions characteristic of occasions on which the method is useable. I further contend that systematic violation of presuppositions does render a method unusable, in that the (supposed) beliefs issuing from the method under such conditions violate coherence standards for the ascription of belief that I have promised to defend in Chapter 9.

A different worry is that with this analysis of normalcy, there are no reliable methods. For, the preconditions and presuppositions of a method's use do not rule out the possibility of error, whereas reliable methods are inerrant under normal conditions. This may not be a problem in the case of artifactual methods, like the thermometer, that provide the leading idea for normalcy; their reliability is a matter of natural law. But consider reliance upon a reference work, the *Encyclopedia Britannica*. Its credentials as an authoritative, trustworthy resource are not such as to guarantee absolutely the truth of all it says. One might argue that it does not take abnormal conditions, like editorial incompetence or fraud, to introduce error; some error is likely to slip by, however exacting the standards of quality in force. And the likelihood increases for a newspaper, which must be rushed into print, or an expert witness, who is not afforded the opportunity to confirm his testimony with other experts.

Chapter 10 addresses these concerns. In the meantime, I agree that it is difficult to *specify* normal conditions under which a method is inerrant. It is difficult to specify, comprehensively, what is presupposed in using a method, for presuppositions need not be recognized or consciously entertained. And some preconditions for a method's use might be unknown. If they are unknown, their absence is not presupposed. The crucial question is how to decide whether a falsehood that a method delivers establishes the method's unreliability, as I understand this, or is attributable instead to abnormal conditions of its use.

An authoritative reference takes stringent measures to ensure accuracy. 'Ensure' is the operative term; the editors do not aim merely at minimizing the error rate or reducing the risk of error. What is the reaction if a mistake is made? Do they say, "Well look, nobody is perfect; mistakes are bound to occur. Consider the great preponderance of truths over falsehoods that we have printed!"? Do they dismiss the odd error as unavoidable? What I see them doing is investigating how the error went undetected and uncorrected. They want to know what went wrong and how to fix it, so that this does not happen again. They fault the *conditions* that allowed this to happen. Such conditions may never be preventable entirely, but they certainly should be abnormal; unless it is reasonable to assume that such conditions are not in place, something has to change. I shall pursue this example in Chapter 4.

The additional vulnerabilities of newspapers and witnesses are, I think, properly understood as challenges to their reliability. It pays to be discriminating with these sources of information. A reliable method based on common media sources, like testifiers generally, will have to incorporate measures to verify trustworthiness.

Consider a different kind of example. I have trouble with names. I fail to identify people correctly when it is socially incumbent upon me to know who they are. My method, facial recognition and memory, is evidently unreliable. But this is the very method I use successfully with people I know well. How can reliability as the property of a method accommodate variation in the effectiveness of a method's use? It seems arbitrary to rely on a boundary of normalcy, and to decree that the errant uses of my method result from its extension beyond the range of close acquaintance. How close does normalcy require? One wants to say, instead, that within their ranges of appropriate application, methods vary in their reliability depending on the objects to which they are directed.

It is important here to distinguish cases in which identification of persons by appearance issues in a false belief, from cases in which it simply fails to operate, or operates inefficiently. My problem is coming up with the name of someone I recognize, or being unsure that I remember the name correctly and so fearful of using it, or hesitant in my identification. I am ignorant, not deceived. Variation in the efficiency of a method does not require variation in its reliability. If no belief is delivered, no false belief is delivered. Short of delivering false belief, a reliable method may induce varying degrees of *inclination* to believe, varying degrees of *confidence* in the truth of a proposition or of action on the assumption that the proposition is true. All of this reliability permits.

I will say more in Chapters 8 and 10 about variation in the trustworthiness with which a reliable method is directed at different objects within its normal domain of application. For now, I contend that it is implausible to imagine someone mis-remembering names with complete conviction. I am inclined to think that for such a person appearance and memory are not a reliable way to identify anyone, not by name. I will, however, introduce some room for intuitive leverage in the notion of reliability in Chapter 4. In particular, I suspect that the problem, in my case anyway, is best understood, not as the unreliability of a method, but as negligence in its use. The problem is not, unfortunately, that I suffer from diminished capacity to retain names, but rather, more to my discredit, that I don't bother to. A case of genuine impairment is a case of unreliability.

The difficulties I have canvassed in the notion of normalcy are distinctive of my approach to reliability. It is not to be supposed, however, that they represent a special liability for this approach, and can simply be avoided by opting for a frequency in-terpretation. It seems straight-forward that a belief formed in a way that yields truths with high frequency is unjustified if formed under conditions on whose absence this track record depends; at least it is unjustified if the believer knows of or has reason to suspect the occurrence of such conditions and their relevance.

This criticism applies whether frequency is assessed over the actual output of a method or includes counterfactual applications. Production by a method that would yield truths with high frequency would not justify a belief were the conditions of this belief's production to be conditions in which truth would be infrequent. Ac-cordingly, frequentists like Alvin Goldman and Ernest Sosa need to incorporate constraints on the conditions in which beliefs are formed into their accounts of justification. Specifically they must, although they do not, condition the justifiedness of beliefs formed by reliable methods upon the exclusion of exceptions to reliability. They cannot dismiss such exceptions as cases of unreliability without turning high frequency into infallibility and disallowing justified false belief. Nor can they spec-ify the exceptions in purely frequentist terms without the same result. They cannot declare a method of forming beliefs justificatory except under conditions in which its output falls below a certain truth-ratio, for the disjunction of conditions in which false beliefs are yielded constitutes such a condition. They will need a principled basis for identifying situations in which use of the method is not justificatory, a basis not dictated by abstract considerations as to the nature of justification but sensitive to what is responsible for the success of the individual method. The frequentist's obligation in this regard is no less onerous than mine.

3.4 Intentional Belief-Formation

I will say more about normalcy and address other possible concerns about the subjunctive interpretation of reliability, including its reliance on counterfactuals, in Chapters 7 and 9. For now, I will assume that we have a working understanding of what it is for a method of belief-formation to be reliable. Then I can address the status of beliefs that result from reliable methods. I shall say that a belief is produced (or sustained[6]) *reliably*, if the believer produces or sustains it by the intentional use of a reliable method or process, under conditions that do not obviously (to the believer) vitiate this process. That is, the believer uses the method intentionally, the method is reliable, and the believer has no reason to believe that conditions are abnormal. The believer uses the method under the (possibly incorrect) presupposition that conditions in which the method would be insensitive to the belief's falsity do not obtain. The reliability of the method need not be part of what the believer intends; it is not (necessarily) *qua* reliable that he intends to use it.

The restriction to *intended* uses of methods is innovative (possibly for good reason). It is intended to resolve the problem of generality that afflicts reliability theories of justification,[7] but before putting it to this end let me issue and discuss some caveats.

I do not suggest that beliefs are typically voluntary; I doubt that we intend to have them. We do not normally decide what to believe, although we do decide what is true (what the truth is), and, in so doing, form beliefs. Although one does not intend to form the particular belief one forms, it can still make sense to describe one's formation of it as intentional. In picking a card from a deck I do not intend to pick the ace of spades; I can't see what the cards are. But I do intentionally pick that particular card, under a different description (the one in the middle). So too, I can intentionally use a method of forming beliefs and thereby form a particular belief, without intending to form that belief. The opacity of intention makes such descriptions ambiguous. Unambiguously, we intend to learn the truth, to find things out, and beliefs result from acting on such intentions. The belief can be involuntary even if the option not to address the matter at all was open.

Even if a belief is formed unintentionally, it does not follow that the method by which the belief is formed is *used* unintentionally. One may intend to find something

---

[6] I will leave this alternative implicit for convenience, when I judge its omission innocuous.

[7] The problem of generality is pressed, for example, by Earl Conee and Richard Feldman (1998). It was acknowledged by Goldman in (1979) and discussed in his (1986).

out without intending any particular method of inquiry. But in acting on the former intention, one necessarily does adopt some means of inquiry, and this move would seem to be intentional. To the extent that the selection of a method is random or inadvertent, it is unclear how one can be described as acting with the intention to learn anything. To act with this intention one must do what one takes to be informative. This attitude explains why one does what one does, and amenability to such explanation seems sufficient to qualify conduct as intentional.

In saying that one's use of a method of belief-formation is intentional, I do not suggest that it need be deliberate or even conscious. An action one is unaware of performing can be (although it need not be) intentional in virtue of complying with one's wishes or interests. This is not to deny that much belief is formed unintentionally, or that it may be contrary to what one does intend that a belief is formed at all. I will consider such cases in due course. But short of proposing a general analysis of intentional action, when a method is used intentionally I think I can defend the compatibility of its intentional use with the evident nature of belief.

The test of what one intends is what one does, and what one would do if things were different. Replace my copy of *Paul Bocuse* with *Joy of Cooking*, and see if I use it. If I do not trust *Joy of Cooking* where I would have trusted *Paul Bocuse*, then my method of proportioning egg yolks to sugar for crème anglaise is not to consult a cookbook as such, but is more specialized. How specialized? Would I have settled for *Julia Child*? Admittedly, the evidence testing produces will always underdetermine one's intended method. My method could be to use M or M', where M is reliable and M' is not. That is, what I intend is: M or M', maybe depending on something random or incidental; I intend neither M nor M' separately. No amount of testing will eliminate all possible candidates for M'.

Or, I could intend different methods. I might use *Julia Child* in place of *Paul Bocuse*, not because it is my intended method to use a cookbook of a certain kind or quality, but because I intend each of these methods independently and use them both as opportunity arises. If the available evidence does not decide between these cases, discovering what I do when presented both books will afford greater discrimination as to my intended method(s). Maybe my methods are conditional, giving one book priority. I use each when it alone is available, but the same one consistently when both are. But again, there will be uneliminated alternatives. Maybe my method is to choose at random and then stick with that choice, whenever possible, for consistency. The problem is that the imputation of an intended method is a hypothesis whose warrant is necessarily inconclusive. This is no more of a problem for judging reliability than for any ampliative inference. Tests, in general, are not definitive.

3.5 The Problem of Generality

Consider, now, what one intends to do when one forms a belief. I look something up in the *Encyclopedia Britannica* and believe what I read. Whether or not I believe reliably, whether or not what I have done produces my belief reliably, is underdetermined by this description of what I do, for this description is compatible with indefinitely many distinct intentions.

If my intention is to consult, and to trust, this particular work, the *Encyclopedia Britannica*, then I take it that my belief is reliably formed, for this is paradigmatic of an authoritative work; trusting it is a reliable method, as I understand the reliability of methods. If my intention is to consult this particular, numerically distinct book, not the *Encyclopedia Britannica* as such—I would not trust your copy or the library's; I'm attached to my own, the gift of a friend—then something rather odd is going on, something at odds, that is, with pursuit of the epistemic goal. Yet my method is still reliable, and I contend that my belief is reliably formed. My method has an adventitious feature that is undesirable with respect to the epistemic goal, but it is nevertheless reliable. The goal is still being advanced, just not efficiently. If my intention is to consult an authoritative work, and it just happens to be the *Encyclopedia Britannica* that I use—it's the one conveniently to hand—the diagnosis is less clear. Perhaps a work is authoritative and so qualifies for use under the method I intend without being trustworthy on the subject of my inquiry. If it is stipulated that to consult a work authoritative *on this subject* is what I intend, then I grant the reliability of my method and of the formation of my belief by it.

However, if my intention is just to consult a book, even to consult a reference work, then although it is the *Encyclopedia Britannica* that I consult I do not think that my belief is reliably formed. That is, even though I do exactly what I would have done had my intention been such as to produce belief reliably, my belief is not produced reliably. If we look just at what I do, without considering what I intend to do, then whether or not my belief is reliably produced is ambiguous, because I do many things, some of which are reliable and some not.[8] This is half of the generality problem. Whatever I do in forming a belief can be described at a high enough level of generality to lose any claim to reliability. Described with sufficient generality, my method could be carried out by actions that are distinctly non truth-conducive. There are very bad books.

Conversely, if my action is narrowly enough described, its epistemic propitiousness becomes unavoidable. For my action did result in a particular true belief, the (presumptive) truth recorded in the *Encyclopedia Britannica*. If coming to hold *this*

---

[8] And, philosophers who press the generality problem against reliabilism will want to add, some things I do are more reliable than others. But variations in the reliability of processes of belief-formation do not align with variation in the justifiedness of the beliefs formed. A relatively unreliable process can produce highly justified belief, and a relatively reliable process can produce less justified belief. My theory will not be subject to this line of criticism, because reliability as I understand it does not admit of degree. Conee and Feldman (1998) further complain that a single process may produce beliefs widely divergent in their degree of justification. So to glance quickly yields a highly justified belief that there is a tree outside, but a poorly justified belief as to the number of its leaves. Unlike reliability, justifiedness does, on my theory, vary in degree (see Chapter 5). But it is only the property of being justified, not its degree, that depends on the reliability of method. My treatment of variation in degree of justification is independent. The Conee-Feldman objection applies to me only if an otherwise reliable process produces an unjustified belief through application to matters with respect to which it is unreliable. I deal with this possibility in Chapter 10.

numerically distinct belief on *this* occasion is part of what I did, so that not just the semantic content of the belief but also its truth is part of what the method is individuated to include, then it is not possible for me to do what I did without believing truly. My belief is reliably formed because my method, to do what I did, could not have yielded a falsehood. Any method capable of yielding a falsehood would have to be distinct from mine, for mine includes this belief as its outcome and this belief is true. This is the other half of the generality problem. Clearly, the particular belief that results from the application of a method is *not* part of what one intends to do in using the method.[9] If one already knows what to believe, *no* method is applicable; there is nothing to do. So this half of the generality problem, like the first half, does not arise for my version of reliable belief-formation.

Are there other methods that guarantee reliability in the event of truth without definitionally incorporating truth, so that they can be intended? A method that does not identify a particular belief may identify some other numerically distinct event or condition that imparts belief in the case of truth. But I think that such a method will turn out unreliable or unintended. For example, I may decide to believe a proposition, which is in fact true, if a particular flip of a coin yields heads, which it in fact does. But the guarantee of true belief here does not satisfy the counterfactual requirements of reliability, even if the coin is fixed (which would be an abnormal condition anyway). Moreover, that the coin yield heads is not part of what one intends, so that what one intends does not force a true belief.

The problem was supposed to be that the method one uses can be described with varying generality. One's actual course of action unavoidably instantiates methods too general to be reliable. And described narrowly enough to include the resultant belief (or something else that dictates the resultant belief) as a defining feature, one's course of action is unavoidably reliable. Reliability is then incoherent. The solution is that what one intends to do is not subject to indefinite gradations of generality.

Neither are methods, really. Strictly speaking, the generality problem is not generated by describing the same method at different levels of generality, for there is no reason not to let the level of generality individuate methods. Checking an encyclopedia and checking a reference work are different methods. Judging visually and judging perceptually are different methods, for one can be used without the other being used. The problem, rather, is that a single act can employ both methods. Every act employing a method also employs a more general method, so that whether or not one acts reliably is indeterminate. Is consulting a properly functioning thermometer a reliable way to determine temperature? It would seem to be, but in so doing, one looks at the thermometer on an even-numbered day; on a Tuesday; through eyeglasses; through a transparent impediment; through an impediment; after drinking

---

[9] As a consequence, anticipated in Chapter 1, Nozick's accommodation of reliable methods that fail to generate beliefs, although these beliefs are true, is untenable. This accommodation requires the belief produced by a method to be a proposition already identified by another method. But to believe a particular, independently identified proposition is no part of what one intends in applying a method of belief-formation. The use of Nozick's derivative method does not qualify as epistemically probative, on my theory.

coffee; after drinking; after swallowing a vitamin pill; after swallowing a pill. We quickly reach a level of generality that subsumes looking at a thermometer on Bourbon Street during Mardi Gras through a strand of multi-colored beads after four mint juleps and an amphetamine—not a reliable way to determine temperature. Enter intention. If one acts out of the intention to use a particular method, this intention will fix the level of generality at the level of the method intentionally used, regardless of what other methods are (also) used.

It may be true, however, that when one's intention is to consult the *Encyclopedia Britannica*, one *also* intends to consult a book. I do not agree that one need intend the entailments of one's intentions. One might select the encyclopedia under the description *authority* rather than *authoritative text*, and not intend to consult a book. Even so, it is plausible in this case that one intends to consult a book, and I grant the extrapolation if we agree that the intention is not to consult just any book, a book as such. The point, then, is that one does intend to do something the doing of which is not reliable as a means of forming beliefs.

I find this complication innocuous. So long as one intends to use a process that is in fact reliable, it does not matter what *else* one intends; one's belief is reliably formed. That is, if *any* of the methods one intentionally uses is reliable then the resulting belief is formed reliably, even if other methods one is intentionally using are unreliable.

This result is completely general. It does not depend on one's intention to use one of the methods arising in virtue of, or depending upon, one's intention to use the other, as in the case of intending to use the encyclopedia and intending to use a book. A belief based on evidence reliable in my sense—evidence one would not have, under normal conditions, were the belief false—may at the same time serve one's interests, and we may suppose that both reason and wishful thinking operate. It is easier for me to imagine that the subject is willing to use either method, and would intentionally use either were the other not available, than to imagine the subject actually using both methods to form the belief at the same time. No matter; if both methods are intentionally used then a reliable method is intentionally used, and that is enough to render the belief reliably formed.

I concede, however, that whether a reliably formed belief is reliably *sustained* may depend on what unreliable methods were also intended in forming the belief. The involvement of unreliability of methods of formation could subvert the reliability of a process of prolonging conviction. Maybe memory is adversely affected. But we should not expect the epistemic benefits of reliable formation (whatever they turn out to be) to be preserved automatically. Reliable belief could be fleeting.

Nor does it matter that although a reliable method was actually used, an unreliable method would have been used had something been different. Maybe the subject would have used the unreliable method had he been unable to use, or been prevented from using, the reliable one. Maybe the truth-value of the belief affects which method is used, such that were the belief false it would still have been formed, but by a different, unreliable method. Nevertheless, on my analysis the belief is reliably formed.

I do not care whether wishful thinking has the greater hold on the subject's psyche. Nozick (1981, p. 180) and Armstrong (1973, p. 209) worry about cases of overdetermination in which different methods, one satisfying conditions for knowledge and one not, are used, or would have been used, to form the same belief. Is the belief known? Nozick decided that he needed to weight the methods (1981, pp. 182ff.). The belief is known if (to a first approximation) the method that satisfies Nozick's conditions for knowing outweighs, or at least is not outweighed by, the method that violates them. As my theory makes reliability depend on method, it may raise the same worry with respect to reliability. If so, in my theory this worry is immediately dispatched without the complications of weighting. Any belief formed by the intentional use of any reliable method is reliable. As no involvement of further methods affects the belief's reliability, none renews the problem of generality.

If this result seems too stipulative, or convenient, remember my defense of the claim that a stipulative explication of justification is a necessary preliminary to theorizing. I have assumed that epistemic justification advances the epistemic goal of believing truths without believing falsehoods. I have identified the justification of beliefs with their formation by truth-conducive methods. As my analysis of truth-conduciveness is to proceed in terms of reliability, I naturally want an analysis of reliable belief-formation that applies where the epistemic goal is advanced. And the use of reliable methods advances the goal, whether or not unreliable methods are also used.

Consider again the case of two books, and suppose that one is reliable and one is not. I form a belief using the reliable one, but would have used the other were the reliable one unavailable. This information does not distinguish my having two methods that I do or would use intentionally, from my having a single disjunctive or conditional method that I use intentionally. As the single method is unreliable, it is then unclear whether my belief is reliably formed. We can propose further tests, but I think it will have to be stipulated that the matter may never be determined definitively. Does this possibility reintroduce a problem of generality? It does not, so long as there is some fact of the matter as to what my intended method is. We trust the hypothesis that best fits the observed behavior, when that behavior is sufficiently rich to discriminate among hypotheses that background knowledge presents as viable explanations of what a person situated as I am and behaving as I do is up to. The absence of guarantees will not impugn my theory more than another.

How plausible is it that in typical cases of justified belief there is a fact of the matter as to what method one intentionally uses in forming the belief? If you find this implausible, you will not grant me a solution to the problem of generality. But let us be clear that the difference is over whether or not there is, as a matter of empirical fact, the relevant intention. The difference is not over whether recourse to the intention with which one acts in forming beliefs handles the generality problem, not unless you have some other objection. The difference is not over the conceptual resources of my theory, but over the extent of its applicability to real belief-formation.

I share the latter concern, and have said that I do not claim that in forming a justified belief one always, or even usually (how could we quantify?), intends a specific method. I certainly claim that my theory applies to the justification of an important and wide range of ordinary belief, and that it applies systematically within the sciences where reliable methods abound. But even if my theory accommodates them, automatic, involuntary, unreflective perceptual beliefs are poor examples for me; I do better with cookbooks and encyclopedias. In this respect, my theory contrasts with past versions of reliabilism that were directed at epistemic fundamentals and had foundationalist motivations. A quick look at the theories that Goldman (1979) offers his reliabilism as a replacement for will set mine apart. It is not clear to me that ordinary, unreflective perceptual beliefs either have or need justification, and I have no epistemic point to make that depends on their justification. I will, in Chapter 7, offer my services against the skeptic, but my theory is not designed for that competition and its viability does depend on winning it. The scope and limits of my theory's application are assessed in Chapter 10.

Even granting an intentional component to belief-formation, there may be a concern as to whether the believer's intentions have the specificity requisite to identify methods. It is natural to have general intentions as well as, often prior to, specific ones. I might form the intention to take account of a philosophical objection, without intending any specific measure to do so. I have not yet assessed the extent of its ramifications. I have not decided among the courses of refutation, assimilation, and diffusion. My intention, so far at least, is only to address the matter one way or another. This intention is (somewhat) indeterminate. So too, I might intend to assess P, even to learn whether or not P, without intending to employ any specific method of inquiry. Certainly this intention is no basis for justification.

But suppose my intention is to assess P reliably, though not in any specific reliable way. As I understand reliability, this is a second-order intention to intend. For, to come to believe reliably is to come to believe through intentional use of a reliable method. Justification depends upon the reliability of the method one intentionally uses. The second-order intention is not a basis for justification, because its object is not a method but an intention. It is a second-order intention about methods to use rather than a first-order intention to use a method. As reliability is not a method but a property of methods, the intention to believe reliably is not the intention to use a method. On my theory, P's justification will depend in this case on the reliability of the method one forms a first-order intention to use in carrying out the second-order intention. And the reliability of this method is independent of whether the second-order intention is specific as to reliability.

In a concession to the spirit of virtue theory, other aspects of which I shall criticize later (primarily in Chapter 9), I would like to say that intentions to use reliable methods, at both the second and first orders, are intellectually virtuous. My theory does not specifically require this form of virtuousness for justification. For it does not require second-order intentions, and the reliability of the method one intends at the first order need not be part of what one intends. There is, however, a close connection. To believe justifiedly will require, on my theory, good reason to believe one's intended method reliable, and I take this constraint on one's intentions to be a requirement for a form of intellectual virtuousness. Thus, my theory does (will) restrict believing justifiedly to believing through the exercise of virtuous intentions.

I do not, however, go so far as to say that believing justifiedly is believing *out of* intellectually virtuous intention, or through the exercise of an intellectually virtuous faculty. For the virtuousness of one's intention could be incidental to one's formation of it. The reliability that one has good reason to believe one's intended method to have need not be one's motivation for intending it. I do not see a need for additional exigence in a theory of justification aimed at the epistemic goal.

3.6 Implementation Versus Instantiation

The restriction to intended methods makes another large difference. No longer does the particular method one uses on a particular occasion of belief-formation instantiate more general methods and subsume narrower ones. No longer do the relations of instantiation and subsumption that generate the generality problem even hold. Instead, the particular action one takes in forming one's belief *applies* or *implements* or *executes* the method one intends to use. One's action is not a method at all, but the implementation of a method. Methods are not types with actions as tokens. Of course, one's action extends to the formation of a particular belief; that is part of what one does. But the reliability of one's action is not at issue; the action itself is not coherently describable in terms of reliability. Reliability attaches to the method, not to the particular sequence of events that carry it out on a given occasion of its use. This sequence could have unfolded without the application of any method at all. In reaching for the encyclopedia one was merely exercising one's arm; in flipping through its pages, one's fingers; in focusing on a particular entry, one's eyes; in registering a fact, one's brain. Then nothing one did can be assessed for reliability. As methods are inherently general, reapplicable on different occasions generating different beliefs, the second half of the generality problem (the specificity half) does not arise.

Conceptualizing the relation between method and act as intentional implementation, it is not possible to suppose that beliefs imposed upon one by external manipulation beyond one's control are reliably formed, not even if the "manipulator" is Mother Nature imposing sense impressions that automatically instill beliefs, not as I understand reliability. Even if the truth of beliefs that result from manipulation is guaranteed, their formation cannot qualify as reliable. For the believer is not applying any method at all; that is being done by the manipulator.

A deranged neuroscientist disables me and operates on my brain, putting me in a brain state that I could not, nomically, be in without believing myself to be in it, nor believe myself to be in without being in it. We might suppose that beliefs are brain states, and that a certain brain state is identical to the belief that one is in it. Then my beliefs are being formed in such a way that they could not be false, but not by me.[10]

---

[10] Compare Alvin Goldman (1979).

Perhaps it is tendentious to deny that *I* am the one forming them, though I urge *formed in me* over *my* forming them, and *imposition* over *formation*. They are, by hypothesis, *my* beliefs. But I am not doing anything intentionally that produces them. Even if I can be manipulated into intending to implement a method that produces them, I cannot, in this scenario, implement the method. By hypothesis, it is what the manipulator does that produces them.[11]

It is an embarrassment as acute as the generality problem that a reliability theory should qualify such beliefs as justified. This is because although the beliefs cannot help but be true, one cannot help but have them whether the supposed nomic basis of their truth obtains or not. Nor would it suffice for justification that this basis be made essential. For the role of the manipulator is incidental. I could suffer from a naturally occurring neurophysiological condition of which I am unaware, in which, even, I have good reason to disbelieve, but that, with nomic invariance, produces the belief that one is in it.[12] I sympathize with reliability theorists embarrassed by tales of external manipulation and aberrant brain chemistry, and prescribe my account, on which such beliefs are not reliably produced.

I hope, now, to have developed a viable account of what it is for a belief to be reliably formed. We can then consider how reliable belief-formation connects with justification.

---

[11] The weaker challenge of Graeme Forbes's (1984) hypnotist case is met in the same way. Forbes supposes that a hypnotist implants only beliefs he knows to be true. The subject cannot be hypnotizing himself (unless he can come to believe what he already knows, in which case justification is unproblematic), so his beliefs are not reliably formed. I also think that hypnotic implantation of beliefs is subject to the criticism I give of clairvoyance in Chapter 9.

[12] Compare Alvin Plantinga's case of the Epistemically Serendipitous Lesion (1993, e.g., p. 207).

Chapter 4
Justification

4.1 The Theory

I can now give a preliminary statement of my theory. The theory is in two parts:

A. *A belief is epistemically justified if it is reliably produced or sustained and no incompatible belief is epistemically justified.*
B. *A person is epistemically justified in believing a proposition that he has good reason to believe is an epistemically justified belief.*

Part *A* bases justification on reliability. In this respect, the theory is a descendant of Goldman's (1979) theory. But reliability is to be understood subjunctively, according to the analysis of Chapter 3. In this respect, the theory is a descendant of Nozick's (1981) theory. This chapter and the next explain the theory. Chapters 6, 7, 8, and 9 defend the theory against objections that I think *A* and *B* can handle. Chapter 10 presents remaining objections and refines the theory in light of them. *A* and *B* are there amplified into a full and final statement of the theory.

A and *B* raise four immediate questions: (1) Why two parts? (2) What is a good reason to believe that one's belief is reliably produced or sustained? (3) What is a good reason to believe that none of one's justified beliefs is incompatible with a belief that is reliably produced or sustained? (4) Why the second clause of part *A*? I answer the first three questions in this chapter, and the fourth in the next chapter.

There is one point, however, that I should confront immediately. I have assumed in the case of the lottery paradox that it will not strain intuition unreasonably to deny that individual lottery beliefs are justified. In any case, I have promised to defend this way of resolving the lottery paradox. There are other paradoxes in which the individual justifications of mutually incompatible beliefs may seem much more secure. Yet my theory conditions justification on the absence of such conflict. On my theory, reliable production is insufficient to justify a belief that conflicts with a justified belief. So reliable production justifies neither of two incompatible beliefs. It remains open that a belief is justified despite conflicting with a belief that is reliably produced, but joint justification of incompatible beliefs is foreclosed. These points will be developed, but I wish to acknowledge up front that my theory will not grant justification in some cases where it may be intuitive to do so. You are justified in believing that a penniless person is poor, while a person with a billion pennies is rich. You may insist that you are also justified in believing that two persons whose wealth differs by only a penny are either both or neither poor, despite your recognition of the impending conflict.

I am not going to propose a solution to the sorites or other paradoxes with this structure. As a semantic paradox that does not depend essentially on epistemic concepts (unlike the lottery and the preface which do), the sorites does not properly lie within the *problematique* of an epistemological theory. Lest this sound legalistic, I note that any solution that respects the truth-conduciveness of justification will somewhere have to restrict justification unintuitively. Unless and until a solution is accepted that localizes the damage by telling us where this is, I suggest that some hesitation in our epistemic attitudes toward propositions that generate such paradoxes is reasonable (indeed, it is exigent). This is all that the credibility of my theory will require.

4.2 A Versus B

The two parts distinguish justified belief from justifiedly believing, and the justification of a belief from the justification of the believer's holding the belief.[1] According to the theory, the justification of a belief is independent of whether or not the belief is believed justifiedly. A belief may be justified although the believer has no good reason to think so, and the believer may have good reason to think his belief is justified although it is not. Indeed, he may have good reason to think his belief justified *without* thinking this, and so without being mistaken as to the justification of his belief. To believe justifiedly, the believer need have no belief at all as to whether his belief is justified. He need not even believe that he believes. People commonly fail to hold beliefs to which they are epistemically entitled.

Does the division in my theory reflect any pre-analytic intuitions about justification? Well, it reflects mine. It reflects the familiar experience that one can believe correctly for poor reasons or incorrectly for good reasons. More broadly, it reflects the independence of normative evaluations of states of affairs from normative evaluations of the motivations and purposes that produce them. I admit that if the division seems artificial to you, I am poorly positioned to defend its pre-analytic intuitiveness, despairing as I have of identifying a clear, distinctively epistemic sense of justification in ordinary usage to begin with. But the preliminary explication borne of this despair underwrites the division, in that goals are things whose achievement can be misjudged. If that does not work for you, I ask you to consider that epistemic intuitions are often in conflict, as evidenced by the fact that both internalist and

---

[1] There is precedent for this distinction, beginning (I think) with Kent Bach (1985). However, my distinction is independent, and I draw and deploy it differently from all precedents that I know of. In particular, my distinction is entirely externalist, whereas the point of Bach's distinction is to identify an internalist form of justification compatible with externalist reliabilism.

externalist epistemologies enjoy strong advocacy. My theory's distinction between the justification of belief and the justification of believing will have a number of applications, as forecast in Chapter 1. But its primary mission is to accommodate conflicting intuitions about justification, and so to improve upon the intuitive credentials of extant theories.

For a quick indication of how this will be done, reconsider the case of believing supposedly expert testimony. The believer is justified by his reason to defer to the testifier as an appropriately expert, sincere authority. Not only is the believer blameless; his reason gives him a positive justification. If in fact the testifier is not expert or is deceptive, then deference to him is not a reliable method of belief-formation. So the resulting beliefs are not justified, on my theory.

That reliable testimony does not deceive assumes that conditions are normal. Under abnormal conditions, reliable testimony can induce false beliefs. In trusting testimony reasonably judged expert and sincere, one presumes that it is not, despite all evidence to the contrary, interested or coerced. If experts were regularly corrupted or forced into testifying falsely, and managed to do so without revealing their malfeasance to the reasonably prudent observer, the divisions of intellectual labor basic to the social fabric would break down. As per my account of normalcy, I assume that this renders testimony inoperative as a method of belief-formation.

And if the corruption of authority is abnormal—if it is of a nature that the reasonable prudence ingredient in trusting testimony does not reveal, so that it must be presupposed absent when testimony is trusted—then my theory still requires that the justified believer have no indication of it. There cannot, for example, be evidence of a local inducement to manipulate belief by crafting convincing testimony. This is why testimony known to be interested is never reliably believable. Conditions are always abnormal for the method of trusting testimony if the testimony is interested, and we know this; we recognize interest to violate what we assume in using the method. My account of reliable belief-formation prevents a reliably formed belief from being discredited by the information that conditions are abnormal. Such information violates a condition for reliability. It could be a reason to believe false a proposition to whose falsity one's method of belief-formation is counterfactually sensitive under normal conditions. One is not justified in believing what one has reason to believe false, even if one also has reason to believe that under normal conditions one's belief would not have been produced if false.

Thus, I do not propose a reliabilist standard of justification to *replace* an evidentialist standard, on which one's justification requires a reason to believe a proposition true. Rather, the reliability of a belief's formation is supposed to *be* a reason to believe it true. *B*-type justification—the believer's justification—requires this reason. My account of reliable belief-formation protects the reason to believe that reliability provides against pre-emption by a reason to disbelieve. Of course, *A*-type justification does not require the believer to possess reasons and so is not evidentialist.[2]

---

[2] Here, for convenience, I have assimilated reasons to evidence. An evidentialist theory may wish to distinguish these, but I do not think the present point is thereby affected.

Without invoking abnormal conditions, we may hypothesize that a source reasonably trusted is erroneous, and this possibility generates the conflict in intuition that the division in my theory resolves. The justification toward which the reasonableness of trust intuitively inclines us is *B*-type justification. The justification away from which the erroneousness of trust inclines us is *A*-type justification.

It is natural to grant justification if the believer is in a position to argue that his method is reliable, and the cause of its unreliability is unrecognizable (by him). Perhaps a drug, surreptitiously administered, induces hallucinations that to the believer are indistinguishable from veridical experience. Asked to confirm or defend beliefs about his environment, the believer exercises the same scrutiny, conducts the same tests that one normally would to justify confidence. Is the light good? Are there obstructions? Do different modes of sense perception concur? Are perceptions congruent with background knowledge and expectations? All of this we imagine constant between the deluded and ourselves. If he's not justified, how can we be?

But, of course, we *can* justify beliefs about the environment through perception and scrutiny of potential sources of error. Whether immediate, involuntary, unreflective perceptual beliefs are held justifiedly, or whether justification is simply not at issue for such beliefs, surely they can *come* to be held justifiedly in a context in which their epistemic status is questioned and scrutiny becomes appropriate. Of course, in so saying I disregard the skeptical option. But my impending discussion of skepticism (Chapter 7) will sustain this disregard. In theorizing as to the nature of epistemic justification, I take paradigmatic cases of justified belief as data. In such cases, the believer is justified and so are his beliefs. I am justified right now in my ordinary beliefs about my immediate environment. In a case of delusion, my theory grants, as we must, that the victim believes justifiedly, even as it denies that his beliefs are justified.

There is really no need for elaborate skeptical scenarios. It is unfortunately common in the sciences not to recognize, despite best efforts, some systemic flaw that compromises the accuracy of an apparently reliable testing procedure. It is always possible to overlook a variable correlated with the effect one is testing for, and so to misjudge, quite justifiedly, the frequency of the effect in a population by inference from samples undetectably biased. As inference from biased samples is unreliable, the conclusions one justifiedly reaches are unjustified.

Conversely, one might form beliefs by a method that is reliable but to the reliability of which one cannot attest. I suppose that one could believe, even with good evidence, that the method is unreliable and use it anyway, though it is hard for me to make much sense of this scenario.[3] We can imagine a clairvoyant cognizant of the ample grounds for skepticism as to the existence of any such capacity and without any suspicion that he possesses it. He gets all these true beliefs, somehow, never

---

[3] Michael Bergmann (2006, p. 168) thinks that beliefs formed by a reliable method are unjustified if the subject believes the method to be unreliable. This is alright, because beliefs formed by a reliable method are not reliably formed unless the subject intends the method, which intention is difficult to impute to the subject in this case. Of course, as Bergmann does not have my theory's A/B distinction, he may mean to be withholding B-type justification with which I can easily concur.

going wrong. But if he just gets them *somehow*, then he is not intentionally apply-ing a method. It is hard to imagine one forming beliefs by intentionally applying a method that one believes, possibly for good reason, to be untrustworthy, let alone nonexistent. Indeed, it is hard to imagine clairvoyance as a source of *beliefs* at all, in one suspicious of it. One gets impressions, imagery, visions, perhaps; but so long as one remains skeptical, one distrusts these spontaneous mental states as informa-tion. Belief may be involuntary, but so is doubt when one is visited by thoughts incongruous with background knowledge. And what one doubts, one does not believe.[4]

So let me weaken the case. One has simply given no thought to the reliability of the reliable method one intentionally applies; one has no opinion as to its reliability, but uses it anyway out of habit or indoctrination. Even this is strained. Upon con-sideration, one would either assent to its reliability, which seems basis enough to ascribe opinion, or question its use. So, weaker still, suppose that one believes one's reliable method to be reliable, or at least believes this likely, but cannot defend this belief. One irrationally trusts a trustworthy method. From some emotional need, one believes the testimony of an utter stranger who happens to be a true authority. Or, one trusts a book because it is the gift of a friend, and it just happens that the book is authoritative. Then one is not justified in believing one's justified beliefs.

Thus, either form of justification is obtainable in the absence of the other.

4.3 Reasons and Rationality

I am supposing that the emotional need, or the connection to the friend, do not function for the believer as *reasons* to think the respective methods reliable. Asked to justify his reliance on the stranger's testimony or the book, the subject would not cite these causes of his credence; he does not regard them as evidence. What if he did? What if one's belief that one's reliable method is reliable is not irrational in the sense of lacking reasons, but the reasons are not good? This presses intuitions about rationality. Is it enough for rationality that one base one's belief on what one takes to be good reasons, or must they *really* be good reasons? There is surely a dif-ference, although the distinction between a reason's being good and being genuine might just be semantic. What one takes to be a reason need not be. I will count it sufficient for rationality that one take oneself to have reasons and that one take these to be good reasons[5] (if this adds anything); but (compatibly with rationality) these

---

[4] As promised in Chapter 1, clairvoyance is to be subjected to a more sustained critique in Chapter 9.

[5] Maybe a consistency constraint is also necessary. The point is that rationality does not require that one be right to treat something as a reason. A person can be rational but insane, because his reasoning is out of touch with reality.

judgments can be mistaken.[6] Justification (of type *B*) depends on *not* being mistaken about this; it requires that one's reasons for taking one's method to be reliable be good reasons, and not merely taken for such.[7] Thus, one can believe rationally but unjustifiedly.

The difference may be expressed by saying that whether one is rational depends on one's perspective. This dependence is innocuous so long as the notion of a "perspective" implicates no more judgment than is required to contrast good reasons from poor reasons, or reasons from other influences upon belief. Certainly, rationality is relative in a way that justification, on my understanding of it, is not. Unfortunately, Richard Foley (1987, Chapter 3) uses the relativity of rationality to one's perspective to draw an "antireliabilist lesson": Reliably believed propositions that are epistemically rational for us to believe would remain epistemically rational in a world in which their manner of formation were massively but undetectably unreliable; for the substitution of such a world for our own involves no change in perspective. There is no argument here against reliabilism as I understand it. The reliabilism of my theory is perfectly compatible with the deceived believer's epistemic rationality, and with his believing justifiedly. What it denies is the justification of his beliefs. The correct lesson to draw from Foley's example is just that reliabilism is a theory of epistemic justification, not a theory of epistemic rationality.[8]

The possibility of unjustified rational belief breaks with the deontological understanding of justification that I have already disputed. The division in my theory is not a means of combining or reconciling externalist and internalist conceptions of justification. The independence of justified belief from justified believing does not plant a foot in both camps. I mean to accommodate (some) internalist intuitions *without* accepting internalism as a philosophical position. The justification part *B* provides for is not internalist, because a bad reason can look good from one's

---

[6] I do not claim necessity. It is not irrational to believe something for which one does not take oneself to have reasons, if there are reasons (whether one's own or not) to believe that for the belief in question, in the context in question, no reasons are needed.

[7] In saying that one has good reasons to take one's method to be reliable, I do not mean to assume that one *does* take one's method to be reliable. Once again, my theory does not require the justified believer to believe his method reliable, only to have good reason for believing this. It is a dispensable convenience to attribute to the justified believer a belief in his method's reliability.

[8] Some of his remarks suggest that Foley might be willing to let this be the lesson. So, in the Introduction to (1987) he says that the book will "say little about the notion of justification", and in Section 3.1 he concedes that reliabilist philosophers need not be interpreted as offering accounts of epistemically rational belief. But other passages assimilate epistemic justification to epistemic rationality. For example, he says (p. 171) that the traditional analysis of knowledge is "rational true belief", and he characterizes the response to Gettier as the addition of further conditions to "rational true belief" (p. 170). And his survey (Section 3.1) of versions of reliabilism fails to turn up anything *other* than rationality for reliabilism to be a thesis about. The versions he delineates (from Section 2.8) do not include alternatives to rationality, but only alternatives to its epistemic form. Below I propose a hypothesis as to what is going on with Foley.

internal perspective.[9] A good reason must really have evidential force, and not just, however blamelessly, be thought to. How is this achieved?

There are numerous possibilities. Their unifying theme is explanationist. I propose that what makes a reason R to believe a method M reliable a good reason is an explanatory connection between R and the reliability of M. This connection can take different forms.

For example, R can be good because the reliability of M must be assumed to explain why R obtains (not why one possesses R, but why R is the case). Then R is poor (inadequate, not good) if it admits of alternative explanations that do not require M's reliability. A poor reason to believe M reliable is not justificatory because in taking it to be good one fails to consider rival explanations, or to accord them due weight. Of course, this neglect could be blameless; one might sincerely and understandably believe that one has adequately canvassed the possibilities when one has not.

Some good reasons are better than others. So the proposed condition for goodness is not quite that M's reliability be a necessary condition for explaining R, but rather that it be necessary to explain R *plausibly*. R is better the more far-fetched and desperate the attempt to explain it under the condition that M is unreliable becomes. What is the difference between a reason's being poor, and its being good but less so than other reasons? I will take the relevant *better than* relation to hold among good reasons. It does not (for my purposes) relate good reasons to poor ones. A poor reason is (equivalently, I think) a condition that does not deserve to be taken for a reason at all. It can be accounted for without regard to M's reliability.

Correspondingly, one can be more or less justified in one's belief, but no threshold of degree of justification separates believing justifiedly from believing unjustifiedly. The question of how justified one is arises only on the condition that one believes justifiedly to begin with.[10] Thus, I am supposing that there is a difference of kind, not just of degree, as to whether or not R's explanation need appeal to M's reliability. If you do not grant the supposition, you can take my usage as stipulative.

Consider an example. Good reasons to believe that the *Encyclopedia Britannica* is reliable, at least across a wide range of topics[11] (that consulting it and believing

---

[9] Perhaps the reason that Foley fails to consider associating reliability with justification, rather than rationality, is that he thinks that the way for epistemic justification to differ from epistemic rationality is for it to become more internalist. He takes epistemic rationality to be ideally reflective, whereas the reflectiveness incumbent upon the believer is normally less than ideal. Failures of epistemic rationality are not normally blameworthy, and so a lesser standard may be thought sufficient for justification, where justification is identified, internalistically, with blamelessness (pp. 234–235). His point about this kind of justification is that the rationality of one's justified belief might be subverted by just a bit more reflection than it was reasonable to expect one to conduct. I share Foley's resistance to excessive internalism, and appreciate his efforts to distance his view of epistemically propitious belief from blameless belief. But idealized rationality is still too internalist a notion to serve the epistemic goal. Idealized or not, rationality is relative to a perspective whose possible defectiveness it may lack the resources to identify or correct.

[10] Recall from Chapter 1 that degrees of justification are to be treated in Chapters 6 and 10.

[11] The restriction of reliability to a range of topics is elaborated in Chapter 10. Unfortunately, even the *Encyclopedia of Philosophy*, let alone the *Encyclopedia Britannica*, does not meet the standard

what it says on a wide range of topics is a reliable method of forming beliefs), are its reputation within learned circles, the scholarly credentials of its contributors, its selection by important libraries, its longevity. It does not explain these attributes to suppose that the work happens to be in fashion, or that a celebrity is reputed to prefer it, or that it has been found to contain a high proportion of truths over falsehoods. Its advocates are skeptical, disinterested, and discerning. And even a very small number of errors on matters of interest and importance would discredit a work of its kind. The plausible explanation for its repute is that the methods and standards followed in composing it are such as to prevent error, even as depth and rigor are achieved. This explanation does not imply that the work is inerrant. It implies that inaccuracy would have to result from some unforeseen and presumptively absent combination of circumstances that do not normally afflict the composition of reference works. Maybe the company that publishes the *Britannica* was secretly purchased by a media magnate who has planted subliminal advertising in the new edition, and this has not been detected. It would take a condition whose nonoccurrence, while not guaranteed, is a normal presumption of the use of reference works.

On the other hand, it is not unusual for venerable institutions to degrade. Maybe the *Encyclopedia Britannica* is past its prime, and the latest edition is no longer reliable though its stature is not yet affected. Like certain expensive restaurants, it is patronized by habit and the inertia of reputation. Maybe it is used because of the social credit that redounds to the user, not because of intrinsic merit. This is a highly implausible explanation. It is immediately discredited by the presence of competition, by the ready availability of independent checks upon the accuracy of any reference work. If the *Encyclopedia Britannica* were the only well regarded resource available, then that it is well regarded would not be good reason to judge it reliable. A state-controlled press does not deserve its credibility.

Another kind of reason to believe a method reliable, not evident in the encyclopedia case, is that if the method does appear to deliver falsehoods, further use of the *same* method reveals and corrects them. A method may be used with greater or lesser skill. Its resources may not be fully exploited. A judgment is reached prematurely; the method, properly deployed, does not sanction it. That this regularly turns out to be the verdict where it appears that beliefs issuing from M go wrong supports M's reliability. I suppose that any method can be misused or used poorly, and we must distinguish such abuse from unreliability. Unless we judge M reliable, we do not expect it to have an unfailing corrective capacity. So to explain its possession of this capacity, we need to attribute reliability to M.

If two mathematicians, intentionally employing the same method of calculation, get different answers, we do not infer that mathematical calculation is unreliable or

---

of reliability with respect to philosophical propositions generally. There may be not only no reliable references, but no reliable methods of belief-formation at all in philosophy. This possibility is considered in Chapter 11. I suppose that the more disputatious and specialized a topic, the less likely there is to be a resource to which one can reliably defer for information about it.

incapable of justifying theorems. If two chess masters recommend different moves in the same position, we do not conclude that their methods of analysis are different. In every game there is a loser who is *trying* to do the same thing as his opponent. One detective solves the crime; another doesn't. Their evidence, training, and experience are the same. As I have defined reliability, a reliable method is not required to deliver answers;[12] the answers it *does* deliver are required to be true, under normal conditions. Where a method that is intuitively justificatory delivers a wrong answer and we cannot fault the conditions, I defend reliability as a standard of justification by suggesting that the problem is not a failure of reliability but of care or skill or diligence in the use of the method. The criterion for the correctness of this diagnosis is that further or more skillful use of the same method reveals the truth. When a closer look corrects an initial impression, the reliability of visual identification is redeemed.

Consider a poor reason to believe a method reliable. Some believe the Bible inerrant because it is the word of God. The problem here is *not* that the reason is poor (in my sense), but that it is (presumably) mistaken. Under the condition that the Bible *is* the word of God, I suppose we can stipulate as to its truth. Compare this with believing the Pope to be infallible because a priest says so. There are any number of plausible explanations of why the priest would say this, whether or not it is true. The priest is defending the authority of the Church, and thereby his position. The priest believes it himself, by unreliable means. The priest so interprets "Pope" and "infallibility" that the Pope's infallibility is a necessary truth. But then its being true that the Pope is infallible cannot be what explains the priest's assuring you of it; his interpretation explains it. Falsehoods can be taken for necessary truths. It is the availability of viable rivals to the explanation *M*'s reliability gives of *R* that discredits *R* as a reason to believe *M* reliable.

4.4 Truth Versus Reasoned Belief

Falsehoods can be justifiedly believed, so there can be good reasons to believe them. A number of possibilities separate a proposition's truth from there being good reason to believe it. Foremost is the possibility that although good, the reason is false. I see no incongruity in this.[13] The perfect circularity of celestial motions was a good reason to believe that different laws of motion apply on earth and in the heavens. The absence of stellar parallax was a good reason to believe that Copernicus was wrong about the motion of the earth. The energy imbalance in beta decay was a good reason to believe that conservation of energy is statistical. The

---

[12] Moreover, a reliable method is not required to deliver correct answers even when correct answers are available. Chapter 8 explains why.

[13] Others may see an incongruity. Timothy Williamson (2000) identifies knowledge with the evidence that justifies belief. He also thinks that what is known is true. So if reasons are evidential, Williamson cannot accept false, justificatory reasons for belief.

constancy of gravitational acceleration was good reason to equate gravitational and inertial mass. That the wine is a sauvignon blanc is a good reason to expect it to complement shellfish. That the bottle has been mislabeled and the wine is actually viognier does not impeach my reason for selecting it to accompany my *plateau de coquillages*.

Of course, once the mistake is discovered and R's falsity is revealed, R is *no longer* a good reason. But this does not imply that it wasn't one before. R need not be true to function as a reason; it is enough that R be taken for true. Therefore, R need not be true to function as an *explanandum* for M's reliability. M's reliability is needed to explain its corrective capacity—nothing else will do—even if M turns out not to have this capacity and there is, after all, nothing to explain.

If this sounds anomalous, you might prefer a harder line: if R turns out false then it was never a reason, but only reasonably taken for a reason. But then the supposed reason for taking R to be a reason could itself turn out not to have been a reason. That the stars had to be close enough to account for the dynamics of rotation of the celestial spheres was considered reason to believe that the Copernican system required observable stellar parallax. The assumption of maximum stellar distance was wrong, however, so it too was not really a reason after all. Gravitational acceleration is not, in fact, constant, and so its constancy was never a reason to believe that gravitational mass equals inertial mass. When gravitational mass and inertial mass were already believed equal, measurements of gravitational acceleration were also inaccurate and a poor basis for the belief that gravitational acceleration is constant. The measurements of celestial motion that confirmed its circularity could have been, and sometimes were, wrong.

Such examples suggest that to account for the reasonableness of the belief that R was mistakenly supposed to be a reason for will require finding and falling back upon epistemically ever more foundational candidates for reasons. The difficulties in this direction are manifest. Ultimately, we will be forced to revert to thoughts or perceptions as reasons. But the role of thoughts and perceptions in reasoning is problematic; in the typical case we do not form beliefs about them or reason from them. Furthermore, it is widely (and correctly, in my view) held that mental states can bear epistemic relations to beliefs only in so far as mental states too are defeasible. If this view of mental states is correct, they do not provide an ultimate fall-back position where the correctness of reasons is secured.

A further difficulty for the fallback strategy is that what we fall back upon is not, in general, a reason for the original belief, but only, at most, for believing there to have been a reason for this belief. For—to anticipate an argument I shall be making in the next section—the *reason for* relation is not, in general, transitive. If reasons must be true, this failure of transitivity implies that the rationality of a belief does not require any reasons at all for the belief. Judgments of rationality are often robust in the face of new information that impeaches our reasons. We judge today that Copernicanism before Galileo was premature. To continue to regard a belief as having been rationally held, despite there never having been a reason to hold it, is harder for me to make sense of than false reasons. I judge unpromising the prospects for an account of the robustness of judgments of rationality that requires reasons to be true. I reject the harder line.

If not true, need reasons at least be justified? Although in the scientific examples it is plausible to regard taking R to be true as justified, in general we can leave the justifiedness of believing R as a matter distinct from R's status as a reason. Let us be charitable and allow that its being God's word is a good reason to believe the Bible. After all, a person who gives this reason is making more sense than a person who cites its expensive leather binding. Considering what it says, it is hard to see how the Bible can be true if it is not God's word.

Another possibility separating reasoned belief from truth is that some discredited or far-fetched explanation of R is in fact correct. The belief that M is reliable is false, even though its truth is needed to explain R plausibly and R is true. Another possibility is that R has no explanation. Another is that R has been misconceived as a proper *explanandum*, not so much because it is false as because it is eclectic; R combines diverse phenomena whose explanations are distinct and unrelated. The University library stocks the *Encyclopedia Britannica* because of the terms of a bequest to the University. Scholars trust the encyclopedia because they or their friends were commissioned as contributors. Its erroneousness has not been discovered because not very much of it has actually been read. What we abstract as a single phenomenon of professional repute for reliability to explain is instead an amalgamation of independently explained attributes.

As none of these possible separations of reasoned belief from truth requires a fallback to abnormalcy, none protects M's reliability. Each is a way for M to prove unreliable despite the need to posit M's reliability to explain a true R. There are lots of opportunities to believe justifiedly that a method is reliable when it isn't.

I am proposing an explanationist account of what makes R a good reason to believe M reliable. My thesis is only that the goodness of R consists in an explanatory connection between R and M's reliability; I do not dictate how this connection must go. The explanatory connection can be the reverse of what we have seen: R is a good reason to believe M reliable because the truth of R is needed to explain why M is reliable. Admittedly, once in search of an explanation of M's reliability one will not (any longer) *need* a reason to believe M reliable. But reasons can be good though unneeded. R can be a good and unknown (or unused) reason to believe reliable a method known to be reliable on other grounds.

For example, R can be an unrecognized necessary cause of M's reliability. The viscous properties of mercury are a good reason to believe the thermometer reliable as an indicator of temperature, because they ensure the proportionality of the height of the mercury to the temperature. The reliability of thermometers could have led to the discovery of these properties. But the accuracy of a thermometer's readings does not explain the viscous properties of mercury. The explanatory connection here is the opposite of what it was in the case of the encyclopedia.

In cases of causal overdetermination, where R is but one of several sufficient causes of M's reliability, the explanatory connection that makes R a good reason to believe M reliable may be inoperative; M's reliability is explained without it. So again, R can be good, compatibly with the explanationism I propose, without being needed as a reason. I claim only that what makes R good is the presence of an explanatory connection, not the dependence of our understanding upon this connection.

4.5 The Complexity of Reasons

My thesis is weak, but I do not need a stronger one. In particular, I do not need to decide what makes a reason good in general. If we consider the good-reason relation in general, not restricting belief to the reliability of a method, it is difficult to circumscribe the range of responsible conditions. The connection could be correlative, indicative, predictive, symptomatic, probabilistic. In some cases, maybe just $Pr(H|R) > Pr(H)$, where H is the belief and Pr is objective probability, is enough to make R a good reason for H. But this will not do as a general criterion.[14] To calculate the probabilities requires contrived and idealized conditions that do not apply in ordinary circumstances where the justification of belief is at issue. To prevent very weak reasons from being good reasons to believe very dubious propositions requires a threshold for $Pr(H)$. And any threshold will authorize good reasons to believe inconsistencies, even recognized inconsistencies. The justification of inconsistent beliefs violates a condition of adequacy from Chapter 1. And—to anticipate discussion of the second clause of A—I shall disallow justifiedly believing recognized inconsistencies. Moreover, I do not think that probabilities are meaningfully assignable to justified beliefs generally, at least not objective probabilities. Unable to prevent people from assigning to propositions the probabilities that they take them to have, I tolerate subjective probabilities; but certainly these are not epistemically justificatory.

Here is an example to illustrate the complexity of reasons. That the keys to the wine cellar are missing is good reason to believe that the butler committed the murder. But the butler's guilt neither causes, nor is caused by, nor explains, nor is explained by, the condition of the keys' being missing. The keys were purloined by the maid with the purpose of deflecting suspicion from the butler. Unable to account for the bottle of *Romanée-Conti* discovered in her chamber, she confessed. To act with her motive, she must have suspected the butler. Not a suspicious person by nature, she must have known something that implicated him. R becomes a reason to suspect the butler by serving as a reason to believe that there is a reason to implicate the butler. The good-reason relation is now quite out of control.

For example, it seems to me that the good-reason relation is not transitive, and in some situations is intransitive. That is, x may be a reason for y and y for z, so that one can reason from x to z, while at the same time x is a reason for $\sim z$. For had the maid been prudent and resisted oenological temptations incidental to her purpose, she would have succeeded; the missing keys would then have exonerated the butler.

---

[14] $Pr(H|R) > Pr(H)$ is the simplest probabilistic standard, and a number of more complicated ones, some depending on Bayes's theorem, are possible. I think that all of them are open to the objections I have in the simplest case. But the most important objection is that reason depends conceptually on an explanatory relationship and cannot be expressed in wholly probabilistic terms.

Here is another example. For his own amusement and in testament to his prowess as a lothario, Henry undertakes to win Fanny's affections by convincing her that he loves her. To this end, he affects the smitten suitor. His resulting behavior is reason to believe he loves Fanny; it convinces Fanny, who has no inkling of its motive.[15] And Henry's motive—his *modus operandi*, at least—is reason to expect him to exhibit such behavior; for example, it creates this expectation in his sister Mary, to whom Henry has confided his project. I suppose that one could reason from the motive to the emotion via the behavior, if upon reaching this intermediate step one forgets its basis. But the motive directly is evidence that Henry does *not* love Fanny. Starting from the motive, one might reason either to "Henry loves Fanny" or to its negation, possibly with equal rationality. If this can be done, then reasoning as such cannot be justificatory, because a proposition and its negation cannot both be justified on the same evidence.

The ambiguity, and more generally the complexity, of reasons are why I need reliability. One might, with some plausibility, just say that a person is epistemically justified if he has good reasons for his belief. Perhaps a way can be found to block systematically ambiguities of the sort just described. Perhaps qualifications can be added to ensure that the person believes *for the reasons he has*, and that his reasons not be undermined by further information. But the real problem would be to explicate reasons. Reliability is more tractable than rationality, even stipulating that reasons are good.

4.6 Evidentialism

I think that the situation is similar for evidentialism, (now stated as) the view that to believe justifiedly is to believe in accordance with one's evidence. Perhaps a way can be found for evidentialism to provide for justifiedly believing one's evidence. Perhaps we can figure out what it is for one's belief to be responsive to one's evidence, as against one's being in possession of evidence for one's belief. Perhaps we can work out some notion of the balance of evidence, or the net weight of one's evidence. The real problem is to understand what makes something evidence in the first place, to propose a general analysis of evidence and to do so without presupposing a concept like justification or reliability.

Some discussions of evidentialism simply ignore this problem. For example, Richard Feldman and Earl Conee (1985) propose an evidentialist analysis of justification without saying anything about what makes something evidence. Feldman and Conee even presuppose notions of "fitting" evidence and "strong" evidence. This lacuna is especially glaring, as their arguments against rival positions depend

---

[15] Here I depart from the story the better to make my case. Jane Austen's heroine in *Mansfield Park* is never deceived.

on borderline and disputatious applications of the concept of evidence. For example, they take the information that someone versed in one's subject dissents from one's belief about this subject to be evidence against one's belief, irrespective of the basis or content of the dissent. And they take mental states to be evidence, thereby assuming that evidence justifying empirical beliefs about the world does not have to involve or depend upon physical interaction with the world. This assumption invites skepticism and misrepresents how the concept of evidence operates. It has long been contested, going back, for example, to John Austin in *Sense and Sensibilia* (1962), but they do not to defend it.

Feldman himself, in (1988), documents numerous examples of the failure to say what evidence is, though his complaint about them is their failure to address the *possession* of evidence rather than their failure to address the nature of the evidence possessed. His own account of the former—that one possesses only the evidence one is currently thinking of—does not help much with the latter, as it presumes that an adequate notion of evidence is already available. The notion of evidence itself Feldman deliberately leaves wide open.

Other discussions avoid the complexity of the problem by treating evidence indiscriminately. Williamson (2000) thinks that one's evidence is everything that one knows. Ram Neta (2003) thinks that evidence includes whatever one is entitled to assume in context. But evidence must be evidence *for* something; it must bear a relevance relation to a point of inquiry. I know, or can assume, that today is Wednesday, but if this is not evidence for anything then it is not evidence. The detective investigating the crime scene is looking *for* evidence. Williamson and Neta must wonder what his problem is; isn't he surrounded by evidence? Why can't he just look *at* it? A viable account of evidence cannot be so indiscriminate. It must capture the evidential relation. The accounts of Williamson and Neta do not differentiate evidence from mere data. Most data are not evidence, because they do not support any significant inference.

Michael Bergmann (2006) thinks that evidence is whatever indications make the truth of a proposition evident. A proposition is justified for one when it is evident to one that the proposition is true, and one is justified in the measure of one's evidence. But what about evidence that is not evident? What about indications of truth that one misses or dismisses?

If Neta and Williamson are too nebulous in their conception of evidence, Bergmann is too narrow. As Bergmann understands evidence there must first be a specified proposition, and then one identifies as evidence input to which believing this proposition is a properly induced doxastic response. But it can be clear that data are evidentially important without its being clear what specific proposition they support. Evidence must bear the evidential relation to something, but the something does not have to be a specific belief; it can just be an issue or question or problem to which it informs and influences a response.

These complications render evidence a less tractable notion than reliability. On the other hand, reliability is unlikely to be the whole story of justification. *A* and *B*, as formulated, state sufficient conditions only. There can be good reasons without reliability, if only because there can be good reasons without any intentional application of a method of belief-formation.[16] But then, good reasons, even given a general explication of the concept, could not be the whole story either. One does not need any reason in order that one's belief be justified, as per *A*.

4.7 Defeasibility

There remains the question of one's entitlement to confidence that one's justification of a belief is not undermined or offset by facts independent of one's justification. This is the thrust of the third of the questions my theory raised at the beginning of this chapter. The question is usefully posed in terms of "defeaters", independent information that challenges either one's justification for one's belief or the belief itself directly.[17] My version of reliabilism assumes that the absence of defeaters as such cannot be required for justification; what counts is the absence of *justified* defeaters. If *any* defeating fact, whether recognized, recognizable, or inaccessible, abrogates justification, then only truths will be justifiably believed. For the negation of a false belief will itself be a defeater, and I do not see how to make an exception of this perfectly general result without continuing to admit propositional constructions having the same effect. If the negation is true, other true propositions will entail it and so become defeaters in turn.

What I require is that one not be justified in believing anything that one is justified in believing to be incompatible with a belief one holds justifiedly. By "incompatible" propositions I mean propositions inconsistent, not necessarily formally, but in the sense that the truth of one is good reason to deny the other. This relation may hold for informal reasons, and may fail to hold despite formal inconsistency.[18] I take it that one cannot justifiedly believe each of two propositions that one justifiedly believes incompatible, because one's justification for each would defeat one's justification for the other. In particular, reliable belief production is insufficient for justification, because an incompatible belief could also be reliably produced. Suppose that the *Oxford English Dictionary* disagrees with the *Encyclopedia Britannica* on some point, not because either is unreliable but because some abnormal condition produced an error in one of them. Consulting both works on the point of contention does not justify me in believing a contradiction. Rather I am *not* justified in holding a belief that I would have been justified in holding had I consulted just one of these sources.

---

[16] Nor am I prepared to deny that an unreliable method could be justificatory. I am not sure that the obviousness of the impossibility that one's belief is false cannot be justificatory, even if it is possible for a false belief to possess this obviousness. More on this point comes in Chapter 10.

[17] John Pollock (1984) calls these forms of defeat "undercutting defeaters" and "rebutting defeaters", respectively. According to Pollock (1986, p. 39, note 9), the distinction originates with his own (1970), where they are called "excluders".

[18] Chapter 5 elaborates on this difference.

B says that justification requires good reason to believe that no such conflict among justified beliefs exists. Why not require only that there be no such conflict, rather than that one have reason to believe there to be none? I think this alternative is both too strong and too weak. It is too strong because there could be a conflict that one is not positioned to recognize; perhaps it is beyond one's intellectual capacity to recognize it. Perhaps one justifiably reaches two conclusions whose inconsistency is indiscernible. I do not want the result that one's acceptance of these conclusions is unjustified, for the methods or standards used in (presumptively) establishing them may be stipulated to be as exacting as one pleases. The absence of conflict is too weak because one could have good reason to think there is conflict where there isn't. One's perception of incompatibility among beliefs ought to affect one's justification for holding them, even if this perception is mistaken.

Without the distinction between justified belief and justified believing, the epistemic assessment of inconsistent beliefs is problematic. Since it may be undetectable, inconsistency should not pre-empt justification. But as inconsistent beliefs commit one to error, their authorization is not truth-conducive. My theory resolves this tension by disallowing the justification of inconsistent beliefs, while permitting them to be justifiedly believed if there is good reason not to believe them inconsistent.[19] This result implements the constraint that contradictions not be justified, while respecting the requirement that justification advance the epistemic goal.

The additional onus of possessing a good reason to believe that justified beliefs do not conflict does not seem to me burdensome. I take it that the *absence* of any reason to believe that any of one's beliefs undermines one's reliably produced belief, together with the *presence* of a reason to believe this belief reliably produced, constitutes the required good reason. That I have good reason to believe that my belief, *P*, is reliable, and see no incompatibility between *P* and anything else I believe, provides the reason *B*-type justification requires. Indeed, under these conditions it seems to me that *P* itself is reason to deny that propositions incompatible with it are justifiedly believable. Of course, this reason that *P* itself provides is defeatable by the provision of a (yet undiscerned) justification for an incompatible proposition. But then this emergent justification does not justify the incompatible proposition, for it in turn is defeated by the justification for *P* that it defeats. A defeated justification does not justify.

4.8 Absent Evidence

It might seem incongruous to treat the *absence* of a reason as justificatory, for it could reflect epistemic malfeasance. One sees no incompatibility because one doesn't bother to look, or, worse, averts one's gaze at the hint of intellectual distur-

---

[19] Not all the tension is resolved, of course. I acknowledged at the beginning of this chapter that my theory does not resolve paradoxes in which inconsistency among apparently justified beliefs is blatant.

bance. One believes that abortion is murder; one also (invariably) believes in capital punishment. Questioned as to the obvious implication, one opposes the execution of doctors who perform abortions and of the expectant women who get them, but does not waver in one's earlier stands. One acknowledges no difficulty. In this (realistic[20]) scenario, we are reluctant to grant justification to any of the beliefs involved. Does justification not require some honest effort to think through the implications of one's beliefs and to consider negative evidence?

Intuitions evidently diverge on this matter.[21] Suppose that one would dismiss or ignore negative evidence if there were any, but there isn't. Can one not still believe justifiedly on the basis of overwhelming positive evidence? Suppose the falsity of P would be indiscernible; if P were false there could be no evidence of this. But there can be lots of evidence of P's truth. Can't this evidence justify believing P? In contemporary physics there are lots of important hypotheses that cannot (technologically) be severely tested (in a way that they would be likely to fail if false), though they have great explanatory and predictive power and have never failed. Are such hypotheses not justifiedly believable?

I think not, and I offer as a benefit of my theory that it not does confer justification in such situations. Reliable beliefs are beliefs that issue from the intentional application of a method that does not produce or sustain false beliefs under normal conditions. So if one has good reason to believe that P is reliable, then one has good reason to believe that were P false one would *not* have come to or continue to believe P by one's method, assuming normal conditions. One has reason to think one's method *sensitive* to the prospect of P's falsity.[22] There would be indications of P's falsity; there would be indications of conflict with other justified beliefs. This makes the absence of such indications justificatory. Positive evidence without severe testability is not confirmatory, and I do not grant the supposition that though there could be no evidence against P there could yet be good evidence for P. Notwithstanding undiscriminating evidentialist theories to the contrary, conditions that P predicts or explains are not evidence for P unless they would not be predictable or explainable were P false.

Justification does carry requirements of intellectual effort and openness to counterevidence. There may be forms of justification that depend on additional provisions to meet these requirements. Reliability builds them in.

---

[20] Bush the First asserted all three positions within the space of five minutes in a 1988 presidential debate with Michael Dukakis.

[21] I interpret Alvin Plantinga (1996) as in disagreement with what I shall say here.

[22] I bracket, for the time being, the interpretive difficulties that P's necessity would raise. See Chapter 10.

Chapter 5
Inference

5.1 The Transmission of Justification

Truth-preserving inference from reliably formed beliefs is a reliable method of forming beliefs. Beliefs formed by truth-preserving inference from beliefs formed by a method that would not, under normal conditions, have produced them had they been false will not, under normal conditions, be false. For, truth-preserving inference from truths yields truths under all conditions. Thus, truth-preserving inference from reliably formed beliefs satisfies the first clause of condition A of my theory. If the beliefs from which one infers also satisfy the second clause of A and are thereby justified, then the beliefs inferred will also satisfy this clause and be justified. For a justified belief incompatible with an inferred belief must also be incompatible with beliefs that support the inference. Hence, truth-preserving inference from beliefs justified under A yields beliefs justified under A. By condition B, a good reason to believe that one's belief is formed by truth-preserving inference from beliefs justified under A justifies holding the belief. More generally, since justification is truth-conduciveness, truth-preserving inference from justified beliefs is justificatory.

I offer the transmission of justification through truth-preserving inference as a major advantage of my theory over others. Neither of the precursor theories discussed in Chapter 1, for example, offers this advantage. Goldman claims that reasoning is justificatory, but the addition of this claim to his theory generates the lottery paradox. And Nozick has to restrict transmission because his reliability property is not closed under entailment. On Nozick's theory, neither justification nor knowledge has my theory's transmission property. In general, no reliabilist theory of justification or knowledge has provided for the transmission of justification through truth-preserving inference without violating conditions of adequacy adopted in Chapter 1.[1]

---

[1] Again, one might make an exception for Sherrilyn Roush (2005), although her way of "providing" for transmission is simply to assert knowledge of propositions known to be entailed by what is known, without giving this knowledge a theoretical basis.

J. Leplin, *A Theory of Epistemic Justification*, Philosophical Studies Series 112,
DOI 10.1007/978-1-4020-9567-2_5, © Springer Science+Business Media B.V. 2009

To appreciate this advantage of my theory, however, requires precision as to the nature of transmission. It is not that inference takes the justifications of beliefs one infers from and carries *these* justifications over to the belief inferred; beliefs inferred from and beliefs inferred do not, in general, have the *same* justification. Nor is it (just) that the status of *being justified* attaches alike to beliefs inferred from and beliefs inferred. Sameness of justificatory status is a consequence of transmission, but is not equivalent to it. Rather, truth-preserving inference transmits justification from justified beliefs in that such inference is *itself* a source of justification. It justifies beliefs, but does so differently from other reliable methods of belief-formation.

This qualification enables my version of reliabilism to provide for the justificatory effect of reasoning. If we think of reliability simply as the subjunctive property of not being believed if false, then reliability is not generally preserved through deductive inference. I would not believe that there is an apple on the table in front of me if there weren't, but my belief that there is an apple implies that I am not hallucinating (the apple), which I might well believe even if I were hallucinating. Suppose that I use the implication to infer that I am not hallucinating. On my theory, what justifies the inferred belief is not its possession of the simple subjunctive reliability property of not being held if false, but rather the property of having been truth-preservingly inferred from justified beliefs. The inferred belief is justified by a reliability property, but it is the more complex property of having been reliably produced, as defined in Chapter 3. This property is closed under truth-preserving inference. You cannot get beliefs false under normal conditions by truth-preserving inference from beliefs that you would not have gotten had they been false, under normal conditions.

I shall sometimes leave the additional complication of my reliability property implicit for ease of exposition. When I say that a belief or method of forming beliefs is "reliable", I mean what I said these mean in Chapter 3. A method is reliable if it does not yield false beliefs under normal conditions. A belief is reliable if it results from the intentional application of a reliable method. Despite the failure of deductive closure for ordinary reliability properties, reliable belief-formation is deductively closed.

Of course, one may protest that the belief that one is not hallucinating is not, after all, justified. Justification *does* require simple subjunctive reliability, or some other property that is not closed under truth-preserving inference. Maybe justification requires evidence, and evidence for P need not be evidence for P's entailments for these are falsified by conditions that do not affect the evidence for P. Nor, when there is justifying evidence for P, will there necessarily be any *other* evidence that justifies P's entailments. Hence closure fails. The advantage I am claiming for my theory is its capacity to quell such protest.

It will take time to make good on this claim. As a start, recall from Chapter 3 that a belief is reliable if any method intentionally used in forming it is reliable. As a consequence, a belief may be justified even if, were it false, it would still have been formed, albeit by a different method. So the simple subjunctive property of not being held if false is not required for a belief to be justified on my theory. But what if the method by which a justified belief would have been formed if false is unreliable, perhaps flagrantly nonjustificatory? Suppose that what I believe by reason I would have believed by faith were reason to have eluded me. Suppose that what I believe on ample evidence I would have believed by wishful thinking were the evidence unavailable.

There are many examples of this kind, originating (I think) with David Armstrong's (1973, pp. 208–209) discussion of a case proposed by Gregory O'Hair. A father believes his accused son innocent independently of (and prior to) a demonstration of his son's innocence in court. The father's belief is (becomes) justified but would have been (was) held out of paternal partiality. Were the belief false, the father would have held it unjustifiedly. For then there would have been no courtroom vindication, but partiality only. In a similar, widely discussed case, an attorney believes his client innocent on the basis of clearly exculpating evidence, but, being in love with his client, would believe her innocent without evidence.[2]

I provided for such cases in Chapter 3. As the complications they introduce do not there obviate reliability, neither do they here obviate justification. I contend that to deny justification in such cases is to take too hard a line. Justification ought not to be reserved for epistemically flawless believers who use and would use only the best methods. Possibly no believer is epistemically flawless. If it is possible to advance the epistemic goal, it must be possible for epistemically flawed believers to advance it. A belief in fact formed in a way that advances the goal is justified even if it would, were it false, have been formed in a way that does not.[3] It seems to me a subjunctive fact about any justified belief that it might have been unjustified, and would have been under possible suppositions about the believer. And correspondingly, what one believes justifiedly one might have believed unjustifiedly given the right suppositions. These possibilities do not pre-empt justification.

To strengthen the intuitiveness of my interpretation of cases like the attorney case, I propose an analogy to moral justification. The attorney believes what he wants to believe. Must this happy alignment make us doubt his justification? Suppose that, in an emergency, a surgeon must operate without anesthesia, and the patient just happens to be the surgeon's hated neighbor. Suppose that the state prison system has an opening for an executioner. Must it be in the job description that applicants dislike killing? Must the executioner suffer in his work? Isn't it better, all things considered, that he enjoy it? Where's the additional harm? Suppose that the bomber pilot ordered to destroy a nuclear weapons factory is a sadist delighted to find innocent villagers in the way. The policeman enjoys arresting people. Surely

---

[2] These examples are normally presented as challenges to knowledge, whereas I am treating them as purported challenges to justification. Nozick (1981, p. 179) has an example in which a grandmother sees that her grandson is well, but would believe him well on the deliberately misleading testimony of others were he ill. In this example, unlike the former, knowledge is presumed, and the point is that whether one knows must be relativized to one's method. Naturally I would make the same point, with respect to justification.

[3] Note that fragility of normal conditions does not threaten justification, on my theory. If conditions undetectably verge on the abnormal, one's belief could easily have been false, but would nevertheless be justified.

attitude does not pre-empt moral justification in these cases (although something else might). As moral justification for action does not depend on motivation, epistemic justification for belief does not depend on what interest belief incidentally serves.

Now, consider one further step. Suppose the surgeon would have harmed his neighbor had the professional opportunity to inflict pain not arisen. Suppose the executioner would have killed had he not been authorized to do so; he would have become a serial murderer had this outlet not presented itself. The pilot might have become a terrorist had the military rejected him. The policeman might have been a thug; many were. As things are, however, they act entirely within social and legal license. Do these suppositions subvert the moral license, should there otherwise be any, of the actions actually performed? I think not. I wonder how many among us owe their innocence to fortune, and how dependent is society on finding the right people to perform unsavory assignments.

Later Chapters, especially 7 and 9, will further the argument that a simple subjunctive sensitivity to falsity cannot be required for justification. My present concern is more general: *no* property not transmitted through truth-preserving inference can be required for justification. As assumed in Chapter 1, an adequate theory of justification must provide for transmission. If a nontransmitted property is necessary for justification, then justification will violate transmission as well, unless some other property necessary for justification guarantees possession of the nontransmitted property by whatever one truth-preservingly infers. On my theory there is no such further property, nor do I see how there could be on any plausible theory. There is no reason why conditions necessary for justification cannot be formulated independently.[4]

Reasoning, of which truth-preserving inference is a mode, is a principal means of both extending and improving one's belief-system. If reasoning did not preserve justified status, it is difficult to see what value reasoning would have; whereas, manifestly, reasoning is the currency of intellectual life. If, in reasoning to new beliefs and to revisions of existing beliefs, one's conclusions do not inherit the justified status of one's premises, why *reason* to them? Why not invest credence piecemeal, as seems appropriate, unburdened by nonjustificatory inferential connections? Only because reason is justificatory do one's beliefs constitute a *system* at all; that is, only reason makes them systematic. Otherwise, they are but a jumble of disconnected commitments, independently reached. This is absurd on its face, and intolerable to any theory of justification or knowledge that invokes doxastic structure, whether coherentist or foundationalist.

Indeed, that justification be closed under truth-preserving inference is a minimal position; many forms of ampliation should also be capable of transmitting

---

[4] Ted Warfield (2004) has pointed out that failure of closure for a necessary condition for knowledge does not imply failure of closure for knowledge. It is logically open that other conditions guarantee that the known logical consequences of what one knows satisfy the condition. I think that this is *only* logically open; it is not going to happen epistemologically.

justification. Although reasoning per se cannot automatically be justificatory, on pain of authorizing inconsistencies, induction and explanation are crucial to the growth of the sciences, and must, properly used, carry evidential support.[5] If not even truth-preserving inference were justificatory, science, as we know it, would be impossible.

In fundamental science, the discovery of new and unanticipated applications of established laws and principles is often a basis for investing credence, independently of actual or prospective empirical confirmation. One example: Reasoning from Maxwell's electromagnetic theory, Oliver Heaviside discovered in 1888 that the electrical repulsion of charged bodies in uniform relative motion drops off, not by the square of their separation r, as in Coulomb's law, but by $r^2(1 - v^2/c^2)^{1/2}$, where v is the relative velocity of the bodies and c is the velocity of light. Many prominent physicists (G.F. Fitzgerald, G.F.C. Searle) immediately concluded (17 years before special relativity) that velocities greater than c are impossible for charged bodies. They regarded this conclusion not as a hypothesis or conjecture contingent on independent confirmation (of which there was, in any case, no prospect), but as a discovery.[6]

Another example: So secure are the principle of conservation of energy and the second law of thermodynamics that we are justified in believing that black holes radiate, although black holes are unobservable and their radiation unmeasured.[7] From our failure to detect such radiation, we in turn infer the homogeneity of the early universe, a conclusion still more remote empirically. I think such examples show that any epistemology that hopes to make sense of science as a knowledge-acquiring enterprise must grant that inference can be justificatory.[8]

The application of epistemological theory to practice requires that the justification inference is to transmit extend from beliefs to believers. To do justice to how credence is invested in practice, it is insufficient that the inferred consequences of justified beliefs be justified. For, in general, that beliefs *are* justified is not, given the externalism of my theory, a datum ascertainable from practice. That beliefs are consequences of beliefs that there is good reason to believe justified must also be justificatory. Condition B is needed. Specifically, that one's beliefs are reliably formed is a good reason to believe that truth-preserving inference from them is a reliable method of forming beliefs. So if one justifiedly believes the premises of a truth-preserving inference, whether these premises are justified or not, the conclusion of

---

[5] Whether or not we can formulate acceptable general principles of ampliative reasoning, I assume that ampliative reasoning does have proper uses. The very program of inductive logic, viable or not, presupposes this.

[6] I thank Marc Lange for calling my attention to this early anticipation of the relativistic limit on velocities.

[7] I assume that black holes do in fact radiate, as physicists say they do. Of course, this is an optional way to describe the process, for the radiation that a black hole emits does not originate within it.

[8] A potential Bayesian qualification to this requirement is considered later in this chapter.

this inference is also justifiedly believable. Both the justification of belief and the justification of believing satisfy transmission.

Therefore, not only is reliability not a matter of the frequency with which truth is yielded, neither is the cogency of one's reasons for believing a method reliable. Suppose one knows, of *n* methods of belief-formation, that all but one method, which one being unknown, are reliable, for high *n*. That one's method is among these *n* cannot be a good (enough) reason to believe that one's method is reliable. Otherwise, one could justifiedly hold *n* beliefs, each the outcome of a different one of the *n* methods, without being justified in believing their conjunction. I shall argue, to the contrary, that one is justified in believing any conjunction of propositions one justifiedly believes. Believing the conjunction, I shall argue, is a proper application of justificatory inference and must satisfy transmission. Since, as per my condition of adequacy, one cannot be justified in believing a recognized contradiction, this result precludes justifiedly believing all members of a set of propositions that is recognizedly inconsistent.

5.2 Inference and Entailment

Truth-preserving inference is to be distinguished from the logical relation of entailment, most obviously because inference is a psychological act whereas logical relations may be unrecognized, even unrecognizable. Logical relations may also be noninferential, unable to sustain or represent acts of inference. Inference is a process in which what one believes already plays a causal role in inducing one to believe something further. One does not infer a proposition from itself, nor a necessary proposition from a contingent one, nor a contingent proposition from a necessarily false proposition, although these are valid forms of argument. Perhaps we can imagine such inferences occurring, where the form of argument is misconceived. We cannot, after all, rule for psychology. I remember hearing a newscaster declare that *since* the price of meat was going up, meat was becoming more expensive. (Perhaps he intended to distinguish wholesale from retail prices.) I dismiss such cases as exceptional.

More importantly, one may decline to infer what one recognizes one's beliefs logically to entail; nor does the truth-preservation of one's inference require logical entailment. One's response to recognition of entailment may be to suspend or modify existing beliefs, or, where rectification proves elusive, to do nothing; such is the nature of paradox. One does not *have* to believe that one is not hallucinating as a condition of believing justifiedly that there is an apple on the table, even if one recognizes the entailment.

Truth-preservation is not a purely formal relation. It is a modal constraint upon the truth-value of the conclusion to conform to truth in the premises, and this constraint can be of various types. Necessarily, if you start out with truth you end up with it; the necessity may hold for semantic, metaphysical, nomic, or even practical reasons. The world is such that if the premises are true so is the conclusion. There may be all manner of possible situations in which the truth of the premises fails to ensure the truth of the conclusion, but as things are no such possibility can be realized. To adapt an example from Gilbert Harman, champion of the division of inference from logic, if I know that Jones plays guard for the Green Bay Packers and infer that Jones weighs over 200 pounds, my inference is truth-preserving.

Logical entailment necessarily preserves truth only because technical definitions of logical terms are chosen for this effect. As logical terms are used ordinarily, inferences that exhibit the form of entailments may not preserve truth. A person justified in believing that there is no God is not, on this account, justified in believing that if there is then the innocent suffer. If the latter conditional is not a material conditional, it is not logically entailed by the negation of its antecedent. Truth is not generally preserved where entailment depends on truth-functional and other formal properties that misrepresent ordinary reasoning.[9] I have no use for formal entailments that do not justify inference. For my purposes, the notion of entailment can be collapsed into that of a truth-preserving relation that holds out of (some) modal necessity; this, henceforth, is what I shall mean by "entailment".

My point then, to put it roughly, is that (Harman notwithstanding) inference and entailment have this connection: If I believe justifiedly, that my belief entails a further proposition justifies me in inferring this proposition. Of course, I might be unable to draw the inference, because I do not recognize the entailment or because I believe the entailed proposition already. And I need not draw the inference just because I can and am justified in so doing; there could be reasons not to. But if I do, on the basis of the entailment, my justifiedness (though not necessarily my specific justification) carries over. That is, supposing I infer a proposition because it is entailed by an existing belief, either I am thereby justified in believing the inferred proposition or I am not (any longer) justified in maintaining my existing belief. Entailment is justificatory in the sense of being a basis for justificatory inference. And the corresponding principle holds for the justification of beliefs; a belief entailed by a justified belief from which it is inferred is justified.

Crossover principles do not, however, hold in general. That a belief is justified does not guarantee that I am, by inference from it, justified in holding beliefs it entails. Apart from the fact that I may not justifiedly believe my justified beliefs, my inference could proceed through fallacious reasoning. And that I believe justifiedly does not justify beliefs entailed by and inferred from what I believe, although it does justify my believing them.

It will prove convenient and harmless to be a bit loose with these distinctions once the principles are precise. But making them precise requires qualifications difficult to express idiomatically. To obtain a symbolic formulation, I will use "$J(P)$" for "P is justified", where P is a belief (a believed proposition); "$JB(P)$" for "P is believed

---

[9] I intend this point as a concession. For those who advocate the material conditional analysis of the indicative conditional, the concession is unnecessary. My purpose is only to free my theory from this position.

justifiedly"; "$\Rightarrow$" for entailment (in my sense); and "$I(P,Q)$" for "Q is inferred from
P". Then the principles are:

a. If $J(P)$ & $(P \Rightarrow Q)$ & $I(P,Q)$, then $J(Q)$.
b. If $JB(P)$ & $JB(P \Rightarrow Q)$ & $I(P,Q)$, then $JB(Q)$.

b makes clear that the entailment relation is neither necessary nor sufficient for
believing justifiedly by inference. $JB(P \Rightarrow Q)$ is logically independent of $P \Rightarrow Q$.
Q could be inferred on some nonjustificatory ground irrelevant to the entailment,
or inferred justifiedly although the entailment does not in fact hold. $JB(P \Rightarrow Q)$
might hold in virtue of one's having reasoned that $P \Rightarrow Q$, where the mistake
in one's reasoning is, if not undetectable, at least such that normally appropriate
caution and acuity (caution and acuity sufficient for justification) are insufficient to
reveal it. Then one has good reason to believe that Q is reliably formed, whence
$JB(Q)$. There is an important asymmetry between the principles, in that a requires
neither that the entailment be believed nor that it be justified if believed. If $P \Rightarrow Q$
then $I(P,Q)$ is truth-preserving, and truth-preserving inference from justified belief
satisfies my reliabilist condition for justification.

5.3 Inferential License

The remaining ambiguity in what I am proposing is deliberate and ineliminable.
One can invoke the entailment relation to draw the inference, but one needn't; one
is justified if one does, but neither reason nor my epistemic principles compel it.
Whether one takes advantage of the license my principles extend depends, I think,
on standards of credence that cannot be legislated uniformly. Some people tend to
be suspicious, others credulous. Rejecting both sweeping skepticism and indiscriminate credulity as philosophical stances, there remains a range of defensible standards
for investing credence.

From my ordinary perceptual beliefs, I am happy to infer that I am not deluded into believing that there are physical objects about. I have perceptual beliefs that entail that the proximity of physical objects is *not* a delusion. Some
philosophers, though nonskeptics like myself, find this inference precipitous; their
response to recognition of the entailment is to reconsider the justification of ordinary beliefs. Then to recover ordinary justification, these philosophers construct
elaborate lines of argumentation that grant skepticism an intuitive grip I do
not feel.

They become contextualists about justification, for example, granting the justification of perceptual beliefs in ordinary situations but denying it once skeptical possibilities are raised. Then they disallow intercontextual inference as ambiguous. Or
they tinker with the transmission principles. They may reject closure altogether, or
grant exceptions to it. For example, they may disallow transmission where the truth
of what one infers is a precondition of the justification of what one infers from. They

will claim that the justification of ordinary perceptual beliefs presupposes that one is not victimized in a skeptical scenario. Then one cannot justifiedly reject skeptical scenarios by inference from perceptual beliefs.

That one is not systematically deluded by external manipulation is a global constraint on justification, in that it affects virtually any belief, possibly even beliefs about one's own mental states.[10] But the reasoning that blocks transmission for global presuppositions should, if correct, apply to local constraints as well. By this reasoning, inference from beliefs justified by evidence cannot justify entailed beliefs that are presupposed by the evidential status of this evidence. For example, the interpretation of geological data as of ancient origin presupposes that the earth was not created five minutes ago, complete with evidence to the contrary. So, from my justified belief that a chunk of platinum sulfide is two billion years old I may not justifiedly infer that the five-minute hypothesis is wrong.

I reject all these maneuvers, and any others that compromise transmission. Indeed, I do so dismissively. I find it richly ironic that in an effort to avoid skepticism one would compromise, let alone abandon, principles on which the very possibility of science, our paradigm of successful acquisition of knowledge, depends. I do not understand how a philosopher who is supposed to be *refuting* skepticism can accept Humean skepticism as the cost of avoiding Cartesian skepticism. So what do I say to these maneuvers specifically?

Chapter 7 addresses contextualism. Here I will comment that I do not know how, without begging the question, to individuate contexts so as to get the result that whenever the falsity of an inferred proposition is consistent with one's initial justifications, the act of inference changes the context. And I wonder about the inference itself. If it supplants one context by another, then in what context does it take place? Is there a third context, intermediate between those of premises and conclusion, that one temporarily occupies while inferring? Or are inferential relations exempt from contextualism because true in all contexts? But surely (and with the evidence of practice) the justificatory status of an entailment relation on which one's inference is based can be as disputatious as that of the belief one infers. In what context is the permissibility of the inference assessed?

Crispin Wright (1985, 2000) is a major source of argumentation against justificatory inference to the presuppositions of one's justified beliefs. My basic problem with his arguments is that I do not know what sort of presuppositional relation is supposed to hold between scientific evidence and metaphysical possibilities. I assume that "evidence" is a concept to be explicated by successful scientific practice, and it is no part of this practice to grant metaphysical possibilities the power to undercut evidence. I am as prepared, on the basis of the evidence, to dismiss the five-minute hypothesis as I am to dismiss the doctrine of creationism. Stipulating that the fossil record is part of what was created does not, within scientific practice, exempt creationist challenges to evolutionary biology from the reach of evidence. The evidence against creationism is considered overwhelming, regardless of the incorporation into

---

[10] This possibility is explored in Chapter 7.

creationism of explanations of why this would be. Indeed, I claim that the evidential relation obeys the general principle that evidence against a hypothesis is evidence against any stronger hypothesis, including the conjunction of the hypothesis with a disclaimer as to the inability to distinguish the hypothesis from rivals on the basis of evidence.[11]

Wright (1985) claims, on behalf of the skeptic, that evidential relations carry presuppositions that cannot be supported by evidence. The existence of other minds, of a material world, and of the past are his examples. I see why one would think that if these metaphysical propositions are presuppositions of the evidential status of experience, then experience should not be evidence for them. But what is the argument for the antecedent? So far as I can make it out, Wright's argument assumes that evidence requires concomitance: nothing is evidence for anything unless found regularly to accompany that thing in experience (1985, p. 441, note 4). Since we have no access in experience to other minds, to the past, or to material objects, independently of what we take to be experiential indicators of them, it is impossible to establish that they regularly accompany these indicators. We can attest to only one component of the required concomitance. So although what we take for evidence may satisfy the concomitance required of evidence, there can be no evidence that it does. Evidential status must simply be presupposed. And the necessary presuppositions include not only Wright's metaphysical propositions, but also the unattestable covariance of what these posit with their supposed experiential indicators.

This argument takes no account of the evidential weight of explanatory reasoning. The concomitance principle implies that there can be no evidence for the existence of things not known to exist, even if their existence is needed to explain the existence of known things. Unless and until things are found to exist, there can never be evidence of their existence. So things cannot be found to exist on the basis of evidence. And if a hypothetical class of entities turns out to be nonexistent, anything taken for evidence of such entities cannot really have been evidence, however strong it looked.

I find the concomitance requirement arbitrary, unscientific, and implausibly restrictive. There was lots of evidence for the existence of black holes before any were found, if indeed any have been found. If black holes turned out not to exist, we would not decree that there had never been evidence of them. There was evidence in the periodicity of chemical elements for the existence of further elements; that's how gallium, scandium, and germanium came to be discovered. There was lots of evidence for the existence of various categories of elementary particles before they were discovered in cloud-chamber collision experiments. It was *because* of this evidence that we tried to produce and track these particles. The concomitance requirement is difficult to square with the fact that we are frequently wrong for very good reasons, and it makes a mystery of the reasoning by which we frequently come to learn that we are right. It assumes an arid, Humean model of evidence that leaves

---

[11] This principle is not to be associated with Carl Hempel's troublesome "converse consequence condition" (1945), which presumes that the evidential relation is to be understood syntactically.

all empirical commitment beyond solipsism automatically vulnerable to problems of induction and underdetermination. If explanatory inference has any legitimacy at all, as it certainly does in ordinary reasoning and in scientific practice, then Wright's argument is a simple non sequitur.

Wright himself (ultimately) recognizes the need for further argument beyond the concomitance principle. What he supplies is the claim that the phenomena to be explained in skeptical cases are not identifiable independently of the purported explanations that skepticism contests, as they must be "where there is real explanation" (1985, p. 448). It is unclear to me what Wright means to require here, but the readiest interpretation seems to undercut his original argument from concomitance. If we have to apply metaphysical concepts, like the concept of other minds or of objective reality, to identify the experiences we take to indicate the presence of minds or objects, then how can we assume that there actually is experience so identified, evidential or not, unless these concepts have application? If we *do* use experiences in a way that we couldn't use them unless metaphysical concepts have application, then these concepts have application.

Wright seems to be saying that not only does the evidential relation presuppose metaphysics, but also our very practice of taking certain experiences to be evidence presupposes metaphysics. If so, then we have our metaphysics, for there certainly is, as Wright's argument assumes there to be, this practice. If the regularities of experience require a material world, then in assuming there to be such regularities (whose evidential status he then disputes) Wright assumes a material world. Will Wright protest that what evidential practice requires is not that metaphysical propositions be true but only that they be believed, and that metaphysical concepts be used to conceptualize experience? If so, we are back to the original point: the explanation of *this* requirement is that the requisite metaphysical propositions are true. Why is this explanation not evidential?[12]

The pertinent fact about practice is that the disconnection between *explanans* and *explanandum* in successful explanation is not as clean as Wright wants. There is no pure category of data uninfected by explanatory presupposition. Nonetheless, there are explanations. "Real" explanations affect the conceptualization of what they explain, without, thereby, succumbing to debilitating circularity. There is a large literature on this issue, in which the thesis that explanatory power is an epistemic, not merely a pragmatic virtue enjoys substantial advocacy.[13] If explanatory power is evidential, then the metaphysics on which Wright contends that our evidential practices depend can be grounded without the circularity that he alleges.

---

[12] Wright's response to the skeptic is that the metaphysical presuppositions of evidence do not need grounding because they are not factual. But his constraints on factuality arguably disqualify fundamental physics, if explanatory reasoning is not allowed to be evidential.

[13] For some of my contributions and references to others, see my (1997). There I argue, especially in Chapter 5, that explanatory inference must be evidential if ampliative inference in general is to be evidential. That is, nonabductive forms of ampliative inference are not evidentially probative unless abductive forms are probative as well. I will not re-engage this issue here. My principal antagonist on this point is, of course, Bas van Fraassen (1980).

My topic is justification, not evidence. But certainly I can say that the evidential relation is ampliative. Not only does it therefore depend on the legitimacy of abduction (of explanatory inference); it is also subject to defeaters. If we are justified in believing a defeater, then we are not justified in drawing the conclusion the evidence supports. That defeaters do not obtain is presupposed in drawing conclusions from the evidence. This is not, however, a presupposition of the evidential relation itself; otherwise, that relation would be truth-preserving. After all, the five-minute hypothesis, creationism, and their ilk are not *supposed* to show that nothing is evidence for anything. Their threat is systematically to undermine the adequacy of evidence to justify the conclusions it supports. This they purport to accomplish by offering rival accounts of (what they acknowledge to be) the *same evidence*. Were the threat to succeed, these conclusions would not be justified after all, despite the evidence for them, and then there would be no justification for inference to transmit. No need for exceptions to transmission then arises.

I do not wish to deny that evidential status itself ever carries presuppositions. It most plausibly does in cases that are highly content-specific, and the presuppositions are specialized conditions that are independently corroborable. We can verify that power is getting to the instrument, so that a reading of zero is distinguished from no reading at all. We can verify that the watch is real, not a toy. But what if, instead of corroborating a presupposition, one infers it? The justification of presuppositions by inference from beliefs the evidence justifies may appear circular.[14]

This apparent circularity challenges my transmission principles. To assess the challenge, we must be clear what is at stake. This is not an argument about whether the presuppositions of evidence are justified or justifiable. There are many ways in which they may be justified. Even if we include Wright's metaphysical propositions among the presuppositions of evidence, there are many ways in which *these* may be justified. If Wright's metaphysical propositions are in fact true, then (surely) there are in fact reliable methods of coming to believe them and good reasons to believe these methods reliable. We may have to refute skepticism to *prove* that presuppositions are justified, but skepticism need only be false, not refuted, for justification to obtain in fact.

---

[14] It so appeared, evidently, to Fred Dretske (1970), who therefore ruled epistemic operators "semi-penetrating". Dretske's argument against justificatory closure systematically confuses *de re* with *de dicto* justification. I may be justified in believing that a student has been killed without being justified in believing that the student body has been diminished, because I do not justifiedly believe that the person killed was a student. I believe of a student that he has been killed without believing the proposition that a student was killed. Obviously, such a possibility poses no challenge to my transmission principles, which are formulated for believed propositions. On the contrary, such cases recommend the *addition* of corresponding principles for *de re* transmission: If I justifiedly believe of an *x* that it is *F*, then I am justified by inference in believing of it that it is *G*, provided that I am justified in believing *F*'s to be modally constrained to be *G*. And my belief of an *x* that it is *G* is justified if inferred from my justified belief that it is *F*, provided that *F*'s are modally constrained to be *G*. So long as the *de dicto* and *de re* versions of transmission are kept straight, Dretske has no case against closure (not here, anyway; other contributions by Dretske will be considered in due course).

Our question is not, however, whether the (supposed) presuppositions of evidence—be they content-specific or metaphysical—are justified or justifiable. Our question is whether *inference*, specifically, can justify them. Can truth-preserving inference from beliefs justified by evidence justify beliefs that the evidence presupposes? If not, then presuppositional beliefs are exceptions to transmission. I claim that truth-preserving inference from justified beliefs is justificatory. Wright says "no" in the case of presuppositional beliefs. My theory says "yes" without restriction. My argumentative burden is to prove Wright wrong.

I suggest that the appearance of an unacceptable circularity in the inference to presuppositional beliefs reflects a complex of ambiguities. If one must already believe the presuppositions to recognize evidence as justificatory, and so to draw conclusions from it, then these presuppositions cannot now be inferred. For if they cannot now *come* to be believed, they cannot now come to be believed by inference. Where no inference occurs there is no exception to transmission. And if the condition for justification is only that the presuppositions be true, not that they be believed, then inference to them is *not* circular. For if recognition that the truth of a proposition is a precondition for the justification of one's belief does not defeat this justification, then that recognition is a basis for inferring the proposition. Where justification is defeated, again there can be no exception to transmission.

Admittedly, presuppositions may be tacit: not occurrently believed in taking one's evidence to be justificatory, but taken for granted. I take a tacit assumption to be one not actually made but needed, and such that one would be willing to make it upon consideration. The justification of the assumption becomes at issue only upon the exercise of this willingness, and then the original alternatives are re-presented. Inference to a tacit assumption is either impossible, justificatory, or devoid of justification to transmit. In no case is transmission endangered.

I further suspect a second-order confusion of presuppositions of the truth of a belief with presuppositions of its justification. One can infer the former from the belief itself, but the latter only from the second-order belief that the belief is justified. Nothing is paradoxical about inferring that the earth was not created 10,000 years (let alone 5 minutes) ago from the evidence of geology. If it seems paradoxical to infer that the earth was not then created complete with evidence to the contrary from evidence to the contrary, consider that this latter inference depends upon believing not just the evidence of geology but, further, that the evidence of geology justifies believing that the earth is older than 10,000 years. As the age of the earth entails that the earth was not created later together with evidence to the contrary, evidence for the former is evidence for the latter. But no one is going to infer the negation of this conjunction directly from geological evidence. Rather, it will be inferred from the age of the earth, once one believes its age to have been established by the evidence. One reasons through the belief that the age of the earth has been established. And if one believes that the age of the earth has been established, it is not paradoxical to infer that requirements for the justification of the belief that the earth is this old are satisfied.

One can have beliefs justified by one's evidence without having beliefs to the effect that one's evidence justifies these beliefs, and so without being in a position to infer that the presuppositions of evidential justification are satisfied. Unless one is in a position to draw this inference, Wright's objection to transmission is spurious. And if one is in this position, the objection fails because the inference is not circular.

5.4 Inferential Caution

In contrast with my foregoing posture of inferential alacrity, I also find many occasions for reluctance. I am unprepared to infer from my justified belief that it was Jones I saw enter the saloon, that if, unbeknownst to me, Jones has an identical twin then it was *not* the twin whom I saw. Yet, that it was Jones whom I saw entails that it was not any twin of his.[15] Away from home, I do not infer from my justified belief that my house still stands that if, unbeknownst to me, a fire has swept through my neighborhood destroying all houses but one, then the one house spared was mine. I demur at inferring from P a conditional $Q \Rightarrow R$, where P entails R and Q is a defeater for P. One might be less reluctant; confidence in P can entitle one to discount Q and embrace R. I am unsure what to make of a case in which I believe I have established P, and Q is the information that an expert I would normally trust rejects P. I might be willing to infer that the expert errs. So psychologically compelling is mathematical reasoning that were P a theorem whose proof I believe I have understood, I would sooner suspect the expert of confusion or disingenuousness than withhold endorsement of R. I suggest that the relative plausibility of these stances is sensitive both to content and to one's disposition to credulousness. The inferential license I claim for entailment is, accordingly, noncoercive. The reluctance or alacrity with which it is deployed vary with the standards of evidence that one requires one's beliefs to meet.

In cases where the license to infer is unexercised, one may wonder why this omission does not immediately redound to the discredit of the beliefs that support the authorized inference. That one demurs suggests that an implication recognized to be drawable is not drawn. Why not? Won't the reason why not undermine existing justifications? Does my reluctance not create a tension to whose resolution my existing justifications are hostage? How can I consistently uphold my antecedent beliefs while declining to endorse what I recognize their truth to require?

I grant (as, according to the next section of this chapter, I must) that I cannot do so if a defeater for the entailed proposition is causing the difficulty. I suggest that cases in which one demurs are not like this, but rather are cases in which recognition of the entailment raises an uninvestigated possibility of defeat. One may be reluctant to believe by inference a proposition that would be defeated by conditions that one's justifications for existing beliefs do not rule out. But the possibility of a defeater is not itself a defeater; otherwise much paradigmatically justified belief would be

---

[15] I assume that names designate rigidly, so that although the twin is named "Jones", he is not Jones.

unjustified. Mere exposure, so to speak, to a potential for defeat does not defeat, and so one's antecedent justifications are not impaired by the sources of one's reluctance.

Why not drop inferential license altogether, where beliefs inferred from justified beliefs are apparently unjustified? Why not extend inferential caution to the point of obviating the entire problem that closure poses for ordinary justification? After all, this is a way out of paradox that fully respects closure, and it carries a certain plausible realism. For surely I am not alone in declining to believe that the person I saw was not Jones's indistinguishable twin by inference from the belief that it was Jones. Who actually comes to believe that his car has not just been stolen by inference from his belief that the car is where he left it? No one will miss the license to draw such inferences.

But what if someone does? What if someone *insists* on inferring what(ever) he knows the truth of his justified belief to require? How are we to stop him? Do we shout, "Wait! Don't do it! You are creating a paradox!" Is it like restraining the person about to enter a time machine that will take him back to kill his mother before he was conceived? The universe is at stake! I think that if there is a time machine there is already a paradox. Analogously, we cannot legislate for psychology. If there is this justification and this inferential capacity, then there is already a paradox if truth-preserving inference is justificatory. I say grant justification to the inferred belief. There is no better way out.

5.5 Closure

Extending "*P*"'s range from believed propositions to propositions generally, and using "*B(P)*" for "*P* is believed", let us apply my transmission principles to a case in which $P \Rightarrow Q$ and $B(P \Rightarrow Q)$ but neither $J(P \Rightarrow Q)$ nor $JB(P \Rightarrow Q)$. Return to the example of Jones, who plays guard for the Green Bay Packers. Suppose I believe that guards for the Green Bay Packers weigh over 200 pounds, not because of any understanding I have of the physical demands of the professional game but because I've heard that the team's owner is a woman who fancies large men. My inference from Jones's position as guard on the team to his weight is truth-preserving, but not for the reasons I think. Am I still justified in believing that Jones weighs over 200 pounds? Surely not. For one thing, by my reasoning I would draw the same conclusion were Jones a wide receiver rather than a guard. Wide receivers need speed and agility, not mass. But suppose an expert on football tells me that NFL rules stipulate a weight above 200 pounds for guards, and this is the basis for my inference. Now I believe justifiedly, supposing the testifier to have passed reason-able scrutiny, although the inferential basis of my belief is mistaken (there being no such rule). I do not think that, to be justificatory, inference through entailment requires one to assess the modal connection correctly, but it does seem to require that one be justified in believing it to hold. $JB(P \Rightarrow Q)$ is the right condition in *b*; it can hold if $P \Rightarrow Q$ is false, or true for reasons distinct from those that justify believing it.

By contrast, my belief that Jones weighs over 200 pounds is justified on reliabilist grounds even if I mistake the source of the entailment relation in such a way that I am not justified in inferring this belief. The inferred belief is justified if the entailment relation through which one infers holds, whether or not it is justifiedly believed to hold. For the belief is reliably formed. Hence $P \Rightarrow Q$ is the right condition in *a*.

Let me combine principles *a* and *b* and label the combination the *Closure of Justification under Entailment* (CJE). CJE applies both to the justification of beliefs and to the justification of believing them. It says that if *P* entails *Q* and *P* is justified, then if *Q* is inferred from *P*, *Q* is justified. And if *P* is justifiedly believed and is justifiedly believed to entail *Q*, then *Q* is justifiedly believed if inferred from *P* on the basis of the entailment. So far so good, but these results are not yet strong enough to explicate justificatory inferential practice.

Science and, I submit, rationality broadly speaking, require more latitude for the transmission of justification than CJE provides. For our beliefs, taken individually, are relatively impoverished in their entailment relations. Formally, they entail themselves, instances of themselves if they are universal generalizations, existential generalizations of which they are instances, negations of their negations, conjunctions and disjunctions of which they are the sole component, material conditionals of which they are the consequent or whose antecedent they negate, and so on. Beliefs issuing from reliable methods may be structurally complex and richly informative, but we do not advance significantly *beyond* them, we do not significantly *extend* our belief-systems, by inferring what they individually entail. Informally, entailment is richer, because of the ability of background information to generate truth-preserving relations. But to extend a belief-system significantly, the inferential role of additional information will have to be explicit.

We must reason from diverse premises to results not evident in considering the premises individually. In general, propositions have interesting entailments only in concert, because only in concert do they impose a truth-preserving modal constraint upon the conclusion. What we wish to be justified in inferring from our justified beliefs is not, in general, entailed by any of them individually; rather it is entailed by their conjunction. As CJE makes no provision for the justification of conjunctions of justified beliefs, too little justification is available for CJE to transmit.

To appreciate the problem, consider the ambiguity in saying that *P* and $P \supset Q$ entail *Q*.[16] They do so collectively but not individually. What unambiguously entails *Q* is their conjunction $P \mathbin{\&} (P \supset Q)$. But to say that *P* and $P \supset Q$ entail their conjunction presents the same ambiguity. They do not do so individually, and to say that they do so collectively, if it does not mean that the *set* $\{P, P \supset Q\}$ entails the conjunction, is indistinguishable from saying that the conjunction entails itself. It is nevertheless natural to say that *P* and $P \supset Q$ entail *Q*, without mention of the set or the conjunction. If we take this statement literally, its plural grammatical form suggests that the predicate distributes separately over the components of the

---

[16] In speaking of or operating within standard logic, I use "$\supset$" for the material conditional and revert to the formal sense of "entailment".

subject. The suggestion is clearly wrong in this case, as it would be if one said that 2 and 3 are 5. As neither 2 nor 3 is 5, they are not. What is 5 is their sum. On this model, what entails Q is the conjunction, not the conjuncts. Is this the right model for entailment?

Suppose that John and Jim are lifting a table too heavy for one person to lift. If it is true that they are lifting the table but false of each of them individually that he is lifting the table, then we have a rival model on which P and $P \supset Q$ do imply Q. But I think that if they are lifting the table then each of them is, although neither can do so alone. For z to be true of x and y without being true of either x or y individually requires a certain relation between x and y. For example, John and Mary can love one another, although John does not love one another and neither does Mary. Maybe "P and $P \supset Q$ imply Q" can be read on this model, the relation between P and $P \supset Q$ being that one is the antecedent of the other.

I prefer to say that entailing Q is *not* something that P and $P \supset Q$ do but is something that their conjunction does, just as being 5 is not something that 2 and 3 do but is something that their sum does. Admittedly, the addition of Q to a proof sequence containing each of P and $P \supset Q$ as earlier, separate lines is standardly authorized in logical deduction. But this is only because it is standard to authorize the addition of a formula known independently on (certain, selected) truth-functional grounds to be entailed by *conjunctions* of formulas already obtained. It would be pedantic to require that one first obtain the conjunction. Any inference rule for derivations in truth-functional logic stipulates a specific number of preceding steps that must be cited in justifying a new entry; this is the number of conjuncts of the conjunction whose entailment of the new entry is the basis for the rule (allowing, for generality, limiting cases of conjunction with a single conjunct).

The point of denying that (in general) propositions entail what their conjunction entails is to bring out clearly that the closure of justification under entailment embodied in CJE is insufficient to provide for the transmission of justification through inference. Unless conjunctions are justified by the justification of their conjuncts, which do not individually entail them, there is insufficient justification for CJE to transmit.

Accordingly, I prescribe an additional principle, the *Closure of Justification under Conjunction*, (CJC), both for the justification of beliefs and for the justification of believing them. Since entailment authorizes inference, CJC justifies one in believing any conjunction of justifiedly held beliefs. I do not think we need worry about the accessibility of the entailment or the nature of the inference here. In the case of conjunction, one cannot exercise one's right to decline to infer on the grounds I earlier gave for reluctance. A conjunction raises no uninvestigated possibility of defeat not already raised by its conjuncts.[17] If these are justified, so is it. It will be sufficient just to stipulate that one does believe the conjunction of one's beliefs (there being no way to violate CJC otherwise). CJC then says:

---

[17] If you are inclined to dispute this, I ask that you bracket your concern temporarily. We will get to it presently, and then more fully in Chapter 6.

If $J(P_1)\&J(P_2)\&\ldots\&J(P_n)\&B(P_1\&P_2\&\ldots\&P_n)$, then $J(P_1\&P_2\&\ldots\&P_n)$;
and

If $JB(P_1)\&JB(P_2)\&\ldots\&JB(P_n)\&B(P_1\&P_2\&\ldots\&P_n)$, then $JB(P_1\&P_2\&\ldots\&P_n)$.

Why has CJC not heretofore been proposed as a distinct principle of inference? It has been denied, of course; the lottery paradox has even been taken to refute it.[18] Any reliability theory of justification that interprets reliability as a high probability or propensity to yield truths will have to deny CJC or swallow the justification of contradictions. Why has the unacceptability of the first expedient not been as plain as that of the second? Why, if one is to get anywhere inferentially, has the need to *add* CJC to CJE not been appreciated?[19]

On a deontological conception of justification, it is possible that CJC is automatic. One might be unable *not* to believe the conjunction of one's beliefs, in that believing the conjuncts might be *all there is* to believing a conjunction. (In this case the last conjuncts of the antecedents in CJC should be superfluous.) Belief-states need not divide up to correspond with syntax. If one cannot help but believe the conjunction, then one is (presumably) blameless and so justified, if blamelessness is (mis)taken for an epistemic form of justification.

On the other hand, it is also possible that (some) conjunctions of beliefs are *impossible* to believe, because of length, complexity, or the semantic divergence of components. Then (presumably) one could not be justified in believing them because one could not (trivially) believe them justifiedly. One can, however, universally generalize over beliefs to form a second-order belief about beliefs, to the effect that if they are justified so are their conjunctions. Such a generalization is, in effect, an extended conjunction, and CJC justifies believing it. I think this result is unavoidable. It is hopeless to try to restrict CJC to manageability by allowing only so many conjuncts as one's repertoire of entailment relations requires. For the resulting (manageable) conjunctions are in turn conjoinable, and so forth.

On the assumption that contradictions are not justifiable, CJC requires that A (in my theory) preclude, at least, the justification of any proposition inconsistent with a justified belief. By "contradiction", here, I do not mean (only) a proposition of the formal structure $P\&\sim P$, which I will distinguish by the label "patent contradiction", but (more generally) a proposition that is false under all interpretations of

---

[18] I suppose the locus classicus is Henry Kyburg (1961).

[19] Consider the case of Richard Foley (1987). Why does he not appreciate this? According to Foley, it is not legitimate, in general, to reason from epistemically rational beliefs as premises. The premises of reasoning must be "properly basic" beliefs, rather than beliefs made rational by inference from properly basic beliefs. For Foley is concerned that the process of reasoning will reduce epistemic status. But a conjunction of properly basic beliefs will not, in general, be properly basic, on Foley's theory. So according to Foley, it is legitimate to reason from premises from whose conjunction it is illegitimate to reason as a premise. This is backwards. It is only in virtue of the implicative relations of a conjunction that its conjuncts can function collectively as premises.

its structure. This criterion is still formal, so that necessary falsehoods need not be contradictions, but it is broader. I assume that patent contradictions are immediately recognizable as contradictory, and that no one can believe them justifiedly (if at all). But the contradictoriness of contradictions generally may not be recognizable, and they could be justifiedly believed. They could not, however, be justified beliefs. No truth-conducive source of belief, certainly not a reliable source, could deliver them.

I have chosen a stronger formulation of *A* that disallows the justification not only of formal contraries to a justified belief, but also incompatible beliefs. These include propositions whose conjunction with the justified belief would be necessarily false on other than formal grounds. Thus, the second clause of *A* protects the justification of a belief from defeat by incompatible beliefs. Of course such incompatibility need not be recognizable; one may justifiedly believe incompatible propositions.

CJE and CJC are subsumable under a general entailment principle (GEP):

If $J(P_1)\&J(P_2)\&\ldots\&J(P_n)\&I(\{P_1, P_2, \ldots, P_n\}, P)\&$

$\{P_1, P_2, \ldots, P_n\} \vdash P$, then $J(P)$;

and

If $JB(P_1)\&JB(P_2)\&\ldots\&JB(P_n)\&I(\{P_1, P_2, \ldots, P_n\}, P)\&$

$JB(\{P_1, P_2, \ldots, P_n\} \vdash P)$, then $JB(P)$.

(Here "$\vdash$" designates the entailment relation for sets and inference is extended to sets.) If CJE is taken for GEP, then of course CJC will not be formulated as a distinct principle. Perhaps an instance of GEP is intended when one says that a number of propositions entail some further proposition.[20] But GEP is unnatural as a transmission principle for justification, for a set is not a proper object of truth or justification (nor of propositional attitudes generally). And so one leaves it implicit that the set, rather than its members, does the entailing.

There is also a problem for inference from unrecognizedly inconsistent sets. I take it that while an inconsistent set formally implies everything, nothing is (knowingly) inferable from it except in virtue of being inferable from a consistent subset of it. If one does not base credence on what one takes to be false, neither does one base credence on what one takes to be inconsistent. But it may be possible to infer patently contradictory propositions from an inconsistent set without being able to

---

[20] Perhaps not. Some philosophers, e.g. Peter Klein (1976), who speak this way do not in fact accept GEP; for they do not accept CJC, which it subsumes. On the other hand, someone who distinguishes multi-premise closure from single-premise closure, and advocates multi-premise closure as a distinct principle for justification, is accepting GEP. Such a philosopher would be exempt from my complaint that the need for CJC is unappreciated, because GEP provides an alternative way to deliver the transmission that I deliver via CJC. My only problem with such a philosopher is the awkwardness I will note of treating GEP as a principle of justification. The distinction between multi-premise and single-premise closure has recently (2004) been discussed, without advocacy, however, by John Hawthorne.

locate the inconsistency thereby revealed. In this case, GEP denies that all members of the set are justified, although, taken individually, all may seem to be justified. Equivalently, according to CJC, not all conjuncts of a contradictory conjunction can be justified, although all may be justifiedly believed if the contradiction is not patent. Moreover, unless the inconsistency is confinable to a proper subset, no member of the set is justified. For the justification of any member would exempt it from the source of the inconsistency. According to CJC, any conjunction of justified members must be consistent.

As an application, if the detective knows, or justifiedly believes, that the murderer is one of the household staff, but has no evidence to direct suspicion more specifically, then all are suspects in that none can justifiedly be believed innocent. Any defeater of a conjunction whose force cannot be further localized must defeat every conjunct. It does not matter how large the staff is, except that with increasing size it becomes increasingly likely that the evidence will discriminate among its members. The detective who declares that he suspects everyone is not exaggerating.[21]

It also does not matter what justifications the conjuncts have, provided that their strength is uniform. What if each person has an alibi that the detective cannot impeach, and would regard as exculpatory did he not know that someone is guilty? What if in a large lottery every player denies winning? The information that someone has won defeats all this testimony, although such testimony is otherwise justificatory. The statistical information that testifiers sometimes err does not defeat their testimony, but the fact that someone is lying does. It contravenes the epistemic goal by making falsity inevitable.

5.6 Degrees of Justification

CJC effectively dissociates justification from probability. Not only is high probability not justificatory, but also probability cannot measure degree or level of justification. If there were a probabilistic threshold for justification, conjunctions would descend below it in violation of CJC. For, probability rapidly dissipates with extended conjunction, whereas the point of CJC is to preserve justification. The probability of a conjunction is a product of real numbers in the interval [0,1] (either the probabilities of the conjuncts or their conditional probabilities on one another) and will (except in the extreme of necessary propositions) be lower than that of any conjunct. The result of combining CJC with such indefinite diminution of degree of justification would be beliefs that are justified (officially), but (intuitively) are less epistemically secure, less worthy of credence, than beliefs that are unjustified.

How, then, should we provide for the fact that some beliefs are more justified than others, while respecting CJC? I have not yet proposed a theoretical basis for variation in the extent to which beliefs are justified, as I have for justification per se and for variation in the extent to which one is justified in holding a belief. I shall

---

[21] Chapter 6 contends with apparently paradoxical implications of this result.

propose such a basis in Chapter 10. For now, it is enough to note that this deficiency challenges only the completeness of my theory, not its correctness. I acknowledge the (plain) fact that justification admits of degree, and propose to accommodate this fact in the way anticipated in Chapter 4: I restrict the relation *more justified than* to the class of justified beliefs; beliefs outside this class are not *less* justified than beliefs in it; they are simply unjustified.[22] This requires that degree of justification not diminish indefinitely through conjunction (nor any form of justificatory inference).[23] Instead, the degree of justification of a conjunction equals that of its least justified conjunct. Then the degrees of justification of the other conjuncts have no bearing on that of the conjunction.

This works nicely, but is it plausible? Consider these propositions: *P*: My tires will last another 50 miles; *Q*: My tires will last another 500 miles; *R*: My battery will last another 5000 miles. I suppose that under normal conditions the tires and battery of a car have comparable longevity, and that my car is fairly new and in good condition. Grant that I am justified in believing each of *P, Q*, and *R*, but that I am less justified in believing *Q* than *P* and my justification for *R* is lower still. Is it plausible that I am no less justified in believing *Q&R* than *P&R*?

Upon reflection, yes. One argument is an appeal to symmetry. Conjuncts are commitments; each carries epistemic risk, and all it takes is the falsity of one to falsify the whole. *Adding* a conjunct to a conjunction one already believes cannot *reduce* one's risk; it therefore cannot increase one's justification. That the conjunct one adds be highly justified does not affect this result. (Nor, I would contend, does any contribution it makes to the coherence of the set of conjuncts.) The level of justification of the conjunction is unimproved. Symmetrically, adding a conjunct low in justification, provided that its level of justification does not descend below that of existing conjuncts, should not reduce justification. For the risk carried by one's existing commitments is already as great as that incurred by the addition. *P&R* is no more justified than *R*, for the addition of *P* cannot reduce epistemic risk. Symmetrically, *Q&R* is no less justified than *R*, for the risk incurred by *R* is already greater than the risk that *Q* represents. Since *P&R* is at least as justified as *Q&R*, their levels of justification must be equal. If this reasoning is convincing, it should not matter if *R*'s level of justification is equated to *Q*'s. (Let *R* say 500 miles.) The level of justification is simply the minimum of the levels of the conjuncts.

---

[22] Thus I reject the view (of, e.g., Steward Cohen, 2005) that, in general, where a predicate has both comparative and simpliciter forms, satisfaction of its simpliciter form is derivative from the degree to which its comparative form is satisfied. For some predicates, the comparative form is derivative, in that we do not use the predicate comparatively unless it is already satisfied simpliciter. I suggest that not only "justified" but normative predicates generally exhibit this priority. Consider "pretty" or "worthy".

[23] Mathematically, one can allow the degree of justification of conjunctions to diminish indefinitely, while imposing a threshold for justification. By fixing a lower limit to degree of justification that conjunction asymptotically approaches, the degree of justification of a conjunction can be made lower than that of any conjunct. The unintuitive result, though, is that the difference between being justified and being unjustified is vanishingly small, rendering justificatory status gratuitous.

We are not compelled to accept symmetry, and there are certainly objections to consider. It might seem that, apart from necessary truths, the more beliefs one has the more likely one is to believe something false. With continuing investments of credence, this likelihood surpasses any pre-assigned level, regardless of how well justified one's beliefs are individually. The next chapter addresses this concern in connection with the paradox of the preface. For now, I must insist on dissociating the relevant sense of likelihood from justification. The risk that one errs somewhere increases with the number of one's beliefs, but the epistemic risk run by a belief is no greater for its inclusion of commitments as or better justified than commitments it already carries. If this dissociation makes justification look arbitrary or stipulative, remember that epistemic justification is justification that advances the epistemic goal. To advance this goal, one *must* risk error. Therefore, epistemic justification *cannot* be something one automatically loses just by incurring risk. Symmetry seems a good way to accommodate intuitions to CJC, which science, reason, and my theory require.

Moreover, the symmetry principle I am invoking may be regarded as an instance of a more general principle that I would expect to carry broad appeal in evaluative reasoning. This, roughly, is the principle that commendation and condemnation be incurable in equal measure. One's subjection to condemnation should be no greater than one's opportunity for commendation. It is as intolerable that the upper limit of one's potential be to avoid blame as that its lower limit be to forego praise. A system of moral evaluation may fault failure only in the measure that it permits success. Unless it is open to me to do right, I am not liable for doing wrong. Correspondingly, if I can be no better off, epistemically, for conjoining, I should be no worse off either. It's only fair.

There is another reason, apart from my admittedly ambitious symmetry principle, why the addition of conjuncts does not, in general, decrease the justification of the resulting conjunction. If it did, say by amounts inversely proportional to the degrees of justification of the added conjuncts, then conjunctions would, in general, be less justified than any of their conjuncts. But the act of conjoining, of believing a conjunction in addition to believing its conjuncts (supposing these belief-states to differ), cannot increase one's risk of error. For one cannot be mistaken in believing the conjunction without *already* being mistaken in believing some conjunct. As the act of conjoining cannot increase one's epistemic risk, conjunctions should have no lower a level of justification than (the least of) their conjuncts.

CJE requires similar protection. Entailment must preserve not only justification as such, but also its degree. Otherwise justification would rapidly attenuate through inference. Reasoning would be justificatory in inverse relation to its sophistication. This result is unacceptable if successful science is to be a model of epistemic achievement.

5.7 The Bayesian Alternative

Speaking of science, this chapter would be incomplete without acknowledgement of the Bayesian alternative. In arguing that science requires that reasoning transmit justification, I ignore the Bayesian theory of confirmation, which purports to represent scientific reasoning without regard to beliefs or justification. Bayesianism replaces belief with degrees of confidence and justification with probabilities. I agree that it is possible to reason with degrees of confidence and probabilities, but my interest is in reasoning with and justifying beliefs.

Science shares my interest. For reasons widely recognized,[24] Bayesianism is inadequate as a theory of science. To these reasons I here add the criticism that Bayesianism is committed to an implausibly absolute (noncontextual) distinction between evidence and theory. Diverse, unregulatable assignments of subjective probability to hypotheses are supposed to converge under conditionalization on new evidence. Only thereby does Bayesianism provide for the objectivity of science. Convergence requires that evidence be exempt from the subjectivity of confidence measures for hypotheses. Otherwise one gets a regress that leaves the final posterior probabilities of hypotheses unconstrained. The evidence must, in effect, be *believed*. And so it is; normally Bayesians just assume that the evidence is true. Of course, if e is to be evidence for h, Bayesians want $Pr(e)$ to be low, since this is one way to make the posterior probability $Pr(h \mid e)$ high. But $Pr(e)$ here is the "expectedness" of the evidence, its probability independently of its being evidence. Low $Pr(e)$ represents the familiar requirement that new information be unexpected and surprising if it is to carry evidential weight. Once e becomes evidence, its probability goes to 1. Only then are we to conditionalize on e.

It is possible, instead, to assign probabilities $\neq 1$ to the evidence, and then use Richard Jeffrey's (1983, Chapter 11) generalization of conditionalization to adjust one's other probability assignments for conformity to the axioms of probability.[25] But then convergence requires that the probabilities assigned to the evidence be objective. If both the prior probabilities of hypotheses and the probabilities assigned to evidence are subjective, the posterior probabilities of hypotheses may diverge rather than converge. One can secure the objectivity of assignments of probability $\neq 1$ to evidence by conditionalizing these assignments on some body of background information, for example the information available during an historical period before the evidence was learned.[26] But then the problem recurs at the level of the background information. This information cannot in turn be merely subjectively probable if the probabilities based on it are to be objective.

---

[24] First and foremost, I, like many others, can make no sense of the assignment of a probability to a scientific theory.

[25] Jeffrey was concerned with cases in which new experience shifts one's subjective probabilities without there existing any new evidential proposition to conditionalize on. His generalization of conditionalization shows how to assimilate the shift so that the axioms of probability are not violated. There is no guarantee that the different probability functions of different subjects will converge through this process of assimilation.

[26] This historical method has been proposed to solve the problem of "old evidence" that arises for Bayesian confirmation. Evidence known before the inception of a theory is unable to affect its probability, and so by Bayesian standards it cannot confirm the theory, which is contrary to scientific practice. A classic criticism of the historical solution is Clark Glymour's (1980).

In effect, the background information must be believed. The point remains that if Bayesianism is to be more than a theory of individual rationality, if it to be a theory of *science*, then it does not eliminate belief. Some propositions, evidential ones or ones on which evidential ones are conditionalized, are proper objects of belief, while other propositions, theoretical hypotheses, are eligible only for revisable probability assignments.

There is no principled basis within science for drawing this distinction, independently of context. What is evidential in one context can be hypothetical and theoretical in another. A hypothesis can be evidence against which further hypotheses are judged, and evidence can be tested against further evidence.

Timothy Williamson (2000, Chapter 9) has an argument against the contextuality of evidence: Some (supposedly) evidential propositions (phenomenal, in his example) are appropriately cited as evidence for themselves in response to challenge. Therefore a challenge to an evidential proposition does not automatically create a context in which that proposition is no longer evidential. I agree with this conclusion, not for Williamson's reason that some evidence is evidence for itself, but for the reason that some challenges are appropriately ignored rather than met with evidence at all. My reason for upholding the contextuality of evidence is not that evidence is challengeable. It need not be. Evidence is challenged on the basis of independent information. A genuine challenge, capable of generating a context in which what is challenged is not evidence, but becomes itself the object of appraisal, must have a basis in some change in the available, relevant information. I uphold the contextuality of evidence because a proposition in need of support may become information that supports something else. Further information could then reopen it to appraisal. Any evaluation of a scientific hypothesis presupposes other hypotheses that, in other possible contexts, would not be presuppositional but disputatious.

Note that because contextual shifts depend on changes in information, and are not achieved simply by issuing challenges or raising possibilities, the contextuality of evidence, as I understand it, is without bearing on skepticism. In particular, it does not imply any relativization of the evidential relation to alternatives relevant in context. It is consistent with (my) contextualism about evidence that P is evidence simpliciter for P_1, not merely evidence for P_1 as against P_2, but has this status in one context and not another. Thus by the contextuality of evidence I do not mean what, for example, Ram Neta means in (2003). Neta contends that an evidential relation can hold only within a context of appraisal that fixes the relevance of alternatives, and that contexts of appraisal are changed by raising hypotheses. I contend that the evidential relation holds on the basis of a relevant body of information, and contextualism is the consequence of the variability of information.[27]

It is unclear in general how firmly or finally a dispute over a hypothesis must be settled to make the hypothesis available for presuppositional status with respect to yet more theoretical areas of inquiry. Certainly, that a hypothesis be *known* is not a condition for it to function as evidence. Williamson's equation of evidence

---

[27] Contextualism as a response to skepticism is considered in Chapter 7.

with knowledge is incompatible with evaluative practice in the sciences. It does not help that Williamson denies infallible access to one's knowledge. The potential to misidentify knowledge does not match up with the potential for shifts in evidential status. A hypothesis used as evidence loses this status if it turns out to be mistakenly believed. It does not lose this status if it turns out only to be believed for good reason but not known.[28] The requirement that evidence be known, accessibly so or not, is unacceptably foundational. More generally, it requires an unacceptable foundationalism to insist upon a stock of privileged propositions unconditionally (context independently) affirmable and able to arbitrate among others. The attribution of this status to any proposition carries potentially contestable theoretical presuppositions.

Moreover, in recognizing even a category of propositions as proper objects of belief simpliciter, Bayesianism concedes its incapacity to treat reasoning and justification comprehensively. The Bayesian needs a theory of epistemic justification to explain the justification of what he takes to be evidence. My theory is available.

5.8 Default and Challenge

I wish to venture a comparison, on this point, between Bayesianism and the "default-and-challenge" model of justification, which purports to offer an alternative to the traditions of foundationalist and coherentist theories. On this model, beliefs carry an automatic presumption of justification, and a question as to *what* justifies them arises only in response to specific objections or challenges that enjoy independent motivation. (The skeptic's error is that he does not grant the presumption, but claims open license to object indiscriminately.) Similarly, with (the prevailing, subjectivist version of) Bayesianism prior probabilities are free, and epistemic obligation arises only in response to new evidence or new recognition of logical relations among hypotheses to which probabilities have already been assigned. In both systems, what one gives a theory of is *response* to outside pressure, never initial epistemic status. The appeal of this priority, it seems to me, is to illuminate epistemic practice.

The practice of giving and getting justifications exhibits a default-and-challenge structure because one cannot question everything at once, and what it is appropriate *not* to question, to take for granted, to assume in argument, varies with context.

---

[28] Williamson (2000, p. 201) says that an unknown justified true belief cannot be part of one's evidence because its falsity is compatible with one's evidence. If one truly believes that draw $n + 1$ yields a red ball, and believes this because draws $1, \ldots, n$ did, that $n + 1$ yields red is not part of one's evidence because it is unknown. This reasoning ignores the contextuality of evidence. In a context in which one is projecting to the next case, the next case is not evidence. In a different context, the next case, or a universal generalization over all cases, could be evidence without any addition to one's knowledge beyond the first n cases. Williamson also contends that evidence must be robust against defeaters, which knowledge is but unknown justified true belief is not. But his argument for this difference assumes that belief comes in degrees, and that even a condition of "outright" belief is compatible with doubt. I am conceiving of belief categorically, as unqualified endorsement. On this conception, the contrast does not go through. Justified categorical belief is no more readily undermined in Williamson's examples than is knowledge.

Thus the default-and-challenge model captures the contextuality of epistemic practice. Similarly, while it requires evidential status to be noncontextual, Bayesianism reflects the instability and diversity of scientific interests in the license it grants to assign prior probabilities. Scientists concerned with different problems and schooled in different methods will, in advance of experimental test, assess new hypotheses differently. Such differences of method or interest may be thought of as contextual differences.

What neither theory recognizes is that the contextuality of practice does not imply that initial epistemic status, in context, either lacks justification or requires none. That it is appropriate in practice *not to request* justification, or inappropriate to request it, does not imply that there need not be any, nor that there need be no theory of it. For even if justification required evidence, grounds, reasons, reliable sources of belief, foundations—whatever, it would *still* be necessary, *in practice*, to presume justification for lots of judgments and impracticable to insist that its basis always be delineated. A theory of when it is appropriate to challenge and defend beliefs is incomplete as a theory of justification, because there is more to justification than justificatory practice. Beliefs can be justified, and can need to be, even if it never becomes appropriate to challenge them, nor necessary to defend them. To grant beliefs a presumption of justificatory status no more obviates the question of what justifies them than a presumption of innocence obviates the question whether the accused is really innocent. These are things that pursuit of the epistemic goal requires us to learn, regardless of their pertinence to practice, whose axiology is far more complicated.

Unfortunately, it is not clear even that beliefs *are* entitled, in practice, to the presumption of justification that it is often pragmatically necessary to grant them. Bayesianism is supposed to apply to scientific hypotheses, and the assessments of scientists may be presumed to reflect some standard of rationality. But the default-and-challenge model is supposed to apply to beliefs and believers generally. Given what we know about what people find persuasive, about how unscientific belief-systems typically are, it is simply untenable to take justification for the default position. Is the (mere) fact that someone (identified only as S) believes P evidence for P? Is it any reason *at all* to incline toward P over $\sim P$? If not, then why suppose that either S or P is justified? In the general case, other things being equal, it makes sense to me to regard belief as incurring epistemic burden. And so it does not seem reasonable to me to treat the default-and-challenge model of justification as the default position in epistemology.

Chapter 6
Epistemic Paradox

6.1 The Preface

According to the paradox of the preface, one is justified in *dis*believing the conjunction of one's beliefs.[1] CJC cannot allow this (supposing one believes justifiedly), but what is to prevent it? One's own experience, like that of anyone else, cautions against confidence in one's belief-system as a whole. It would seem to be justified, indeed compelled on pain of hubris, to expect some among one's present beliefs to turn out false. This expectation is not localizable; it impugns no belief in particular. Hence one's (presumed) justifications for one's beliefs, taken individually, are unaffected by it. But the justification that CJC supplies for their conjunction is defeated.

Two possibilities are consistent with CJC, neither immediately attractive (hence the paradox). One is that the reason the preface paradox gives me for expecting that not all of my beliefs will survive *does* impugn them all. If it is not localizable, it defeats all of my justifications. What is this reason? A modesty induced by experience? A reluctance to make an exception of myself, when others, my intellectual superiors included, have gone wrong? A recognition of human cognitive limitations? Let us grant that extended, comprehensive belief-systems, such as mine is at present, have regularly contained error. My reason to forecast failure within my present system is an induction from this record. One possibility is that this induction supports a very general skepticism.

Another possibility is that the induction fails. The reason is no good; my reluctance to stand up for my beliefs, unwarranted. Worse, the deliberate ambiguity of my intellectual stance, maintaining my convictions individually while backing away from them collectively, is unbecoming. In investing credence, I assume responsibility for the conjunction; to renege on this responsibility while continuing to invest credence is disreputable. I like this option much better than the first.

---

[1] In original form, the paradox arises from the presumptively warranted admission, by an author in the preface to his book, that despite his diligent confirmation of the book's assertions errors undoubtedly remain. These, he adds, are the fault of no one but he who now assures us of his diligence. This seems a disingenuous acceptance of responsibility. If modesty compels it, how much more than modesty is behind the admission that requires it?

J. Leplin, *A Theory of Epistemic Justification*, Philosophical Studies Series 112, DOI 10.1007/978-1-4020-9567-2_6, © Springer Science+Business Media B.V. 2009

Notice how strange the inductive reason is. It is supposed to justify believing the negation of a conjunction without providing any reason or evidence against any conjunct. For any conjunct differentially disparaged could simply be suspended, obviating the problem. Indeed, the inductive basis has no *relevance* to any conjunct. The previous failings from which one induces are presumably *un*related to the belief-system the induction impugns, for those failings will have been (may be assumed to have been) corrected for. The system adjusts to expel error. So I am supposed to believe that some of my present beliefs are false for reasons that have nothing to do with any of them. And these reasons are supposed to prevail over any justifications, however strong, however individually unimpeachable, that my present beliefs enjoy. I don't think so.

6.2 Second-Order Evidence

You can always eliminate at least one suspect in any murder case on *Perry Mason*; Mason's client is innocent. Suppose that a juror knows of Mason's track record and, by induction, discounts present evidence against the defendant. I think that more is wrong with this than violation of the judge's admonition to consider only admissible evidence. The point of the judge's admonition is to keep the trial fair and protect the defendant's civil rights. This is to be done even if the cost is so to impoverish the evidence-base as to prevent determination of the truth. Our legal system systematically sacrifices truth to the protection of civil liberties; the judge's admonition is not truth-conducive. The juror's important error is epistemic; it is to suppose that the induction on Mason's record is a better indicator of the truth than present evidence.

The issue here is the status of second-order evidence, evidence bearing on the trustworthiness of evidence. Second-order evidence bears upon beliefs in virtue of their status *as beliefs*, independently of their semantic or propositional content. The faultiness of previous belief-systems is proffered as evidence against one's present belief-system, regardless of its content. But this induction is surely incomplete; if not content, there must be some other relevant similarity between the faulty systems and the faulted one that subjects the latter to results obtained from the former. Abstractly, that certain elements of a class have a property is insufficient reason to ascribe that property to further elements. We need a connection to project the property within the class. Otherwise, from the fact that every serve I've hit in tennis has gone long I could infer that the next will go long, and that it is therefore pointless to practice my serve.

The evident connection for belief-systems is the methods or standards for holding beliefs. As the content of beliefs is immaterial to the induction, it must be the investment of credence as such that the induction impugns, pursuant to the checkered history of investing credence. As my methods and standards have licensed false beliefs, they are judged untrustworthy. My present system is therefore suspect. But this induction assumes that methods and standards do not change, or that they do not improve; or at least that we do not improve in the effectiveness with which we use them. They do not become more reliable or more successful, as my serve improves with practice. It assumes that what we learn from our mistakes is only to correct the mistakes, never to increase our precautions against making them. I see no reason to grant such a restriction on what is learned.

But even if the restriction is correct, there is another problem. If the methods and standards for holding my present beliefs are no better than those known to license falsehoods, should I not conclude that I am wrong to invest credence as I do? Isn't the correct conclusion not just that my present belief-system contains errors, but that its contents are unjustified? If so, there is no paradox; there is no violation of CJC to contend with, for its antecedent is unsatisfied.

Indeed, the challenge of the supposed paradox seems to me better directed against the justifications of my present beliefs than against their truth. It is more plausible to induce from past error that I am wrong to believe all of my present beliefs justified, than to believe that, *although* all are justified, not all are true. And if I do not believe that all my beliefs are justified, neither do I believe that their conjunction is justified. Perhaps a deontological conception of justification conceals the intuitiveness of this resolution of the paradox. On an externalist conception, it is more credible that one is mistaken, despite one's best efforts, as to the justificatory status of one's beliefs.

Thus my theory of justification offers a ready resolution to the preface paradox. To the extent that the induction from past error threatens my present belief-system, it threatens my justifications for my present beliefs as much as their truth. For it provides reason to distrust the reliability of my methods of investing credence. And if I presently hold beliefs unjustifiedly, then CJC does not require the justification that the paradox purports to defeat.

Although it is adequate as a defense of CJC, I dislike this resolution of the paradox because I find its concessions excessive. I think that the paradox is resolvable independently of my theory. For I find the whole strategy of using second-order evidence to thwart the verdict of first-order evidence dubious. It is, after all, first-order evidence that provides the conclusions that form the inductive basis for one's second-order inference. Unless we trust first-order evidence, we are not entitled to the premise that belief-systems have regularly proved faulty. But at the same time, if our standards for investing credence are systematically lax, then we are wrong to give first-order evidence this trust. So if the conclusion of the paradoxical induction is correct, then its inductive basis is insufficient to draw it. And if the conclusion is incorrect, then of course it is a mistake to draw it. Either way, the proffered case against my present belief-system is unsuccessful.

The only way I see to reinstate the paradox is to allege an asymmetry between justification and refutation. The first-order evidence that second-order evidence impugns is evidence that justifies my present beliefs, not the evidence needed for an inductive basis of past failures. There is no second-order evidence against the reliability of refutation.

But in fact, there is. Is it not a common experience to change one's mind prematurely, only to find that one was right after all? Maybe we should induce on

this experience to the precipitousness of allowing past mistakes to impugn present credence. Induction is troublesome enough in principle. It is far from clear what, if anything, is correctly inducible from so roughly delineated and ambiguous a database as the paradox of the preface depends upon. Are there not large areas of subject matter within which my beliefs have been quite stable? Surely I can learn where to be guarded without compromising my belief-system as a whole. Talk of one's "belief-system" as a whole is in any case vague. How can one draw any conclusion *about* one's belief-system? Wouldn't this conclusion be *part* of the system, and hence subject to any deficiency it itself alleges? Perhaps we should talk at most of *systems*, and abandon uniformity of epistemic stance across beliefs as such, as unrealistic, even incoherent. If so, it becomes unclear how exactly the induction is supposed to work.

I am also doubtful that a principled asymmetry between justification and refutation can be sustained. The program originates with Karl Popper, who had, finally, to admit that it amounted to procedural policy rather than a real difference in the nature of warrant. Refutation is no more definitive than justification, because the refutation of any proposition depends upon the justification of others. We *elect* to respect refutations and to decline the option of protecting favored propositions. This preference measures intellectual integrity, according to Popper (who nevertheless seized just this option to protect his own demarcationist program from historical refutation). But I see no intellectual integrity in upholding beliefs individually while retreating from their conjunction. And I see no epistemic difference that would privilege the evidence and reasoning by which we learn we have erred over that by which we learn to begin with.

If it is hubris to hold that all conjunctions of justified beliefs are justified, then I am afraid that hubris is endemic to philosophical method. The paradox of the preface no more threatens CJC than any philosophical thesis is threatened by the extensive and unremitting record of philosophical error. There would seem to loom the same inductive threat. But it would be outrageous to reject a philosophical thesis against which one is unable to formulate the slightest criticism, on the grounds that *as a philosophical thesis* it is inherently unworthy of credence. Such an objection is obviously self-refuting.

6.3 The Lottery

CJC confronts a second paradox, that of the lottery. This paradox does not arise for my theory, as I do not adopt a frequency interpretation of reliability and do not allow high probability to be justificatory. But the lottery is supposed to *be* paradoxical just because these choices are supposed to be foreclosed. The lottery has been thought to refute CJC on the grounds that high probability *must* be justificatory. The reasoning is that otherwise skepticism is unavoidable. Ordinary, paradigmatically justified beliefs are not certain. They have probabilities below 1, and therefore below the probability assigned to the individual beliefs that a ticket will lose in a large enough lottery. But a justified belief cannot be *less* probable than an unjustified belief. Hence, if the lottery beliefs are not justified, neither are beliefs that we take to be paradigmatic of justification.

I deny, however, that ordinary, paradigmatically justified beliefs have probabilities below 1; at any rate, I deny that their possession of such probabilities, however close to 1, is the source of their justification. That is, if a justified belief *does* have a high probability (which I do not, in general, concede) this is not the source of its justification. Of course, I grant that justified beliefs are not (in general) certain. They are (typically) defeasible. I contend that the lottery paradox confuses the defeasibility of a proposition with the possession by its negation of a positive probability. In general, ordinary beliefs do not have probabilities; neither, therefore, do their negations.[2]

6.4 Justification and High Probability

There are few sources of probability, and the only source applicable to much ordinary belief is statistical. But we do not have a statistical basis for much of what we justifiedly believe, and I suggest that careful consideration of the statistics we *do* have would often *undermine* beliefs normally assumed to be justified. I believe that I am in good health. I feel fine, had a favorable medical evaluation recently, have no history of sudden illness, and do not engage in risky behavior. My belief, I submit, is justified. But I have no statistics as to the frequency with which people enjoying these attributes turn out to be ill. It is not unprecedented to hear of a case of serious illness befalling the unsuspecting. It can happen suddenly to a person with a history of perfect health. Routine medical tests sometimes miss illnesses; they are not definitive, and it is not routine to test for everything. Some illnesses cannot be protected against and produce no symptoms until well advanced. This is why people are advised to get medical checkups, even when nothing seems wrong. Now that I reflect upon the possibilities, I'm not sure how well justified my belief is after all. Maybe I'm sick.

Lots of ordinary belief is like this. My reasons, that I feel fine and have recently been checked, are justificatory, despite the ways they leave open for me to err. Perhaps it does not take abnormal conditions for me to be ill despite my grounds for believing otherwise. But under normal conditions, an illness will show up under further testing and monitoring for symptoms. Further application of the very method by which I now assess my medical condition favorably will reveal my error. My conviction is rooted in the reliability of medical science, not in the infrequency of illness among people positioned as I am. Maybe an insurance actuary can assign

---

[2] You may find the denial of probabilities to ordinary beliefs overstated. And, indeed, the argument I shall give offers the option of saying instead that ordinary beliefs have too many probabilities. If you prefer this option, fine. The difference will not stop you from reaching my conclusion that the justification of ordinary beliefs is not probabilistic.

a probability to my health, but I can't. And if I could, if I carefully studied the relevant data, I suspect I'd be inclined to hedge my belief that I am healthy into the (different) belief that I am *probably* healthy, even if the data were strongly in my favor. That is, I would be so inclined if I did not know better than to be impressed by statistics.

The reference class with respect to which I obtain a statistical probability for my own health is selected for the big picture, not for me. It is indiscriminate with respect to properties that I take to affect my justification. Are its members by nature even-tempered, optimistic, unencumbered by intellectual confusion as to their responsibilities in life? Are they invested in a false ideology? Do they share my sense of humor? What are their culinary standards? Do they drink enough red wine? Any statistical probability as to my health could easily neglect variables that, while they lack statistical significance over large populations and so are properly ignored in forging the big picture, radically affect me.

If my probability for health is high but it turns out that I am ill, the subsequent discovery of an unusual, unsuspected physical abnormality in me would naturally be taken to vitiate the assigned probability. The correct probability must reflect the discovered condition. Additional information can change the probability assignment *enormously*, and there is no reason to expect convergence to a sustainable assignment with increasing information. Any probability assignment to the individual case is unstable in this way. Regardless of the thoroughness of its statistical basis, it can turn out *wrong*.

Thus, the reason that ordinary beliefs (typically) lack probabilities is not that probabilities are unascertained or difficult to ascertain for them. A probability unascertained is still a probability. It could, on someone(else)'s theory, determine the justificatory status of beliefs, though not of the holding of them. Rather, an ordinary belief lacks probability in that there is, in general, no such thing as *the* correct probability of it. Any body of information that does not determine the truth-value of the belief, and so fix its probability degenerately, is, in principle, improvable.

The problem, as I see it, is that *all* probabilities are, at base, conditional probabilities. It might be thought that we can convert conditional probabilities into categorical probabilities by building in the conditions. The probability that a coin will turn up heads given that it is fair becomes the probability that a fair coin will turn up heads. But there are always further conditions. That an individual atom of U^{235} will decay within 7.1×10^8 years has probability 1/2 only conditionally on the laws of nature. And we cannot here build in the laws of nature without redundancy. We cannot say that the probability that an atom of U^{235} *behaving in accordance with the laws of nature* will decay is 1/2, for it is not possible for U^{235} to violate the laws of nature. And if we try to specify the laws of nature, or state that the laws of nature hold, as part of the proposition to which probability is assigned—if we build the laws in as truth-conditions for this proposition—then the resulting proposition has no probability. There is nothing to give it one.

But if all probabilities are, at base, conditional, then the only sense we can make of the "correct probability" of a belief is that this is the probability determined by the correct conditions. In the case of statistical probabilities, the notion of correctness of conditions is elusive; it makes sense only contextually. It reflects both one's purposes and the state of one's information as to how these purposes are best advanced. Any choice of conditions could turn out incorrect, in that it is not the right choice for one's purposes. A choice ignoring a rare or unknown genetic abnormality could be good enough to set government policy for the distribution of vaccines, but it is not good enough to fix the premiums on my life insurance policy if I am afflicted. And whatever one's purpose, the information on which a probability is properly conditioned can change. If it does, then one was operating with the wrong probability.

So I have difficulty understanding how there can be a fact of the matter as to what the probability of an ordinary, paradigmatically justified belief is, notwithstanding our ignorance of this probability or the defeasibility of our estimate of it. I do not know how there can be a fact of the matter as to what is the right statistical basis, in absolute terms, for the probability of the belief that I am healthy. What should be included among the facts, known or unknown, on which the hypothesis that I am healthy is properly conditioned, independent of the purpose that my probability of health is to serve?

But perhaps there are easier cases. The *San Francisco Chronicle* reports that the Giants beat the Dodgers, and so (with a sigh of relief) I believe it. Is there not a fact of the matter as to the proportion of such reports that have, historically, been in error? Doesn't this fact determine a probability for my belief, based as it is, independently of my knowledge of this probability? My problem with this suggestion is that I do not see how, independent of one's purposes, there can be a fact of the matter as to what reports count as "such reports". Are all baseball games included; all sporting events; all reports of the outcomes of competitions of any kind? At what point do reports get included that the *Chronicle* is more likely to get wrong than it is to get a Giants-Dodgers game wrong? And what about other newspapers, other media? Do their records of accuracy count? I deny that there is such a thing as *the* probability of my belief that the Giants won.

What is the probability that I am correct in believing that my car is where I parked it? With what frequency do parked cars get towed, stolen, vandalized to the point of nonexistence, swallowed up in fissures produced by earthquakes, carried off by tornadoes? How do these frequencies vary with make, model, color, condition, and location of the vehicle; with the time of day, of the month, the year, the visibility of objects within, the nature of such objects, the vehicle's vulnerability to foreclosure, its similarity in appearance to vehicles vulnerable to foreclosure? My belief that my car is where I parked it is justified, but it has either no probability or indefinitely many probabilities among which it makes no sense to identify any one as the correct probability.

But doesn't the justification of my belief that my car is where I parked it generate another lottery paradox? Some cars (in the vicinity; let us be loose about this for the moment) are *not* where they were parked. Parking your car enters you into a perverse kind of lottery in which the "winner" loses his car. If my car belief is justified, then so is every individual belief to the effect that one's car is where one parked it. Then by CJC the belief that no car is missing is justified.[3] But this latter belief has a justified negation and so cannot be justified, not while respecting my conditions of adequacy.

John Hawthorne (2004), pursuant to Jonathan Vogel (1990), discusses this problem from almost every angle.[4] What the discussion misses is the relevance of the difference between the logical probabilities of a real lottery and the statistical probabilities of the car lottery.[5] I deny that the negation of the conjunction that CJC delivers is justified, because I deny that high probability is justificatory. It is perfectly possible that no car is missing; the condition that supposedly defeats the justification of my car belief is only statistically probable. All the individual car beliefs could be true; investing credence this way does not make error inevitable. By contrast, the individual beliefs in the real lottery cannot, logically, all be true. It is a defining feature of the real lottery that some ticket wins, and so the number of tickets fixes the probability of winning for each. Investing credence in the real lottery necessitates error, which subverts the epistemic goal.

One can play many variations of the real and statistical lotteries to try to attenuate this difference. I shall consider some of these in due course, including the case of a lottery that need have no winner. But we already have grounds not to expect justification in such a case. Probability is not justificatory, and the epistemic goal does not tolerate leaving avoidance of error to chance. For now, I maintain that my car belief is justified, but not in virtue of a high statistical probability. Rather, to anticipate, I would not under normal conditions remember (or seem to remember, if "remember" is factive) where my car is if it were not there.[6]

The belief that some car beliefs are false is unjustified, because statistical probability is its only basis. If we change things to stipulate that some car is missing, that this is not merely probable but established, then my car belief is no longer justified. The belief then justified is only the belief that my car is very probably where I parked it (which, however, offers substantial intuitive compensation). But all of this is for the sake of argument. It grants that ordinary justified beliefs, like car beliefs, possess determinate statistical probabilities, whereas I have argued that such probabilities are both contextual and conditional.

---

[3] If, as I maintain, an individual's justification for his car belief is not probabilistic, it may not, unlike his lottery belief, extend to other car beliefs, so that CJC gets no purchase. I ignore, for the sake of argument, this easy way to diffuse the car paradox. At some cost in plausibility, we could position the individual to justify an indefinite number of car-type beliefs.

[4] Hawthorne is concerned with knowledge rather than justification. One (possibly Hawthorne himself—2004, p. 8) might be more reluctant to attribute knowledge than justification to car beliefs (or lottery beliefs). If so, Hawthorne's version of the problem is easier than mine.

[5] Oddly, Hawthorne assumes that the basis of (real) lottery beliefs is statistical, but then says that the disinclination to count them as knowledge is independent of this assumption. Apparently, by "statistical" Hawthorne just means "probabilistic".

[6] If you are worried about how a car can fail to be where it is, identify, as per Bertrand Russell's (1957) reply to Peter Strawson, a location, l, where you seem to remember leaving your car, and make the belief that your car is in l.

What, then, about a statistical probability that is both contextual and conditional? Why can't this probability exist independently of being ascertained and then be a basis for justification? After all, I grant that statistical probabilities are in principle obtainable through empirical research for beliefs justified under my theory. So my theory cannot depend on their nonexistence. Suppose that a justified belief possesses a statistical probability conditional on an apparently comprehensive body of relevant evidence. Even if standards of comprehensiveness and relevance are necessarily inexact, doesn't this probability have to be high? If this probability is lower than the probabilities of unjustified lottery beliefs, is the lottery paradox not reinstated? The paradox is reinstated only if this lower probability, in virtue of (still) being high, justifies the belief. Why doesn't it?

One reason is that justification is not contextual in the way that the notion of *the* correct statistical probability must be contextual. The status of being justified is robust over changes in the purposes for which statistical probabilities are needed. This is an argument for Chapter 7. It arises in connection with contextualism, which is a response to skepticism, the topic of Chapter 7. My present argument is that when such a contextual, conditional probability is assignable, not this probability but the reliability of method is the basis of justification. Even if we know how often newspapers misreport the outcomes of sporting events, when I believe by reading it in the *Chronicle* that the Giants beat the Dodgers, what justifies my belief is not the high statistical probability that the paper is right, but its reliability. The *Chronicle* has reliable sources of information and takes precautions against error. Under normal conditions, it is accurate (at least about Giants games). In trusting it, I presuppose that the editor has not been bribed by someone who bet heavily on the game. A major conspiracy might cause the paper to misreport some things. But with the conspiracy revealed and normalcy restored, the reduced statistical probability of accuracy does not pre-empt justification. The scandal would be treated as an aberrant event, not something to condition one's doxastic reactions to daily headlines.

6.5 Justification and Higher Probability

The probabilities that individual lottery tickets lose are logical rather than statistical. The information on which these probabilities are conditioned is stable in a way that statistical information is not. We do not calculate lottery probabilities on the basis of experience in previous lotteries. The probabilities are independent of that experience, independent of there having been previous lotteries. Depending on how a lottery works, the probabilities may be fixed in advance; in any event, they remain equal, independent of the individual case. As a principle of epistemic symmetry, beliefs that fare alike as to evidence and reason should be objects of the same epistemic attitudes.[7] It is by an exercise of this symmetry principle that the

---

[7] I am trying to state the principle neutrally as among different theories of justification. I think that any theory must respect symmetry, because it is a consequence of the assumption that justification

detective takes the evidence that someone is guilty to cast suspicion uniformly over the household staff, absent anything further to go on. He cannot, consistently with CJC and symmetry, withhold suspicion from anyone while holding that someone is guilty. As probability is the only epistemic property of lottery beliefs, symmetry requires that all of them be justified or none.

It is possible to respond to the lottery by denying symmetry. In particular, an epistemic conservatism is possible, according to which one is justified in sticking with beliefs on an epistemic basis no better than that available to rival beliefs. Symmetry is violated if the contingent doxastic history of the believer is justificatory. I am not going to try to disprove such a position here. I will say only that it seems to me to be an obvious confusion of epistemic with prudential justification. My question, rather, is whether the uniform justification of lottery beliefs that symmetry requires matters in posing the paradox. Perhaps high probability can yet be justificatory, without paradox and without sacrificing CJC, in a different kind of case in which the alternatives have lower probability.

We can invent scenarios that privilege certain beliefs while retaining probability as their sole epistemic property. For example, in a lottery with n tickets, any ticket drawn except t_i wins immediately, but if t_i is drawn then it wins only if a flipped coin comes up heads. Then the probability that one's ticket loses is greater by $1 \div 2n$ with t_i than with any other ticket. Do we want to say that one is justified in believing one's ticket will lose if it is t_i, but not otherwise?

I certainly have no inclination to say this. For those unconvinced, there is a general argument to show that the program of understanding justification in terms of probability cannot be salvaged in this way. For, there is *always* an alternative that does *not* have lower probability. Any proposition of probability less than 1 is logically inconsistent with some proposition of greater probability. CJC and CJE then guarantee that if high probability is justificatory patent contradictions will be justified, in violation of a condition of adequacy.

Assume $Pr(P) < 1$. Then $\lim(1 - (Pr(P) \div 2^n))$ as $n \to \infty = 1$. Select n such that $Pr(P) \leq 1 - (Pr(P) \div 2^n)$. Let $Q_1, \ldots, Q_n$ be propositions probabilistically independent of P and of one another, each with probability 1/2. (These could be propositions reporting the results of n flips of a coin.) There are 2^n distinct Boolean conjunctions of the Q_i; that is, conjunctions the ith conjunct of which is either Q_i or $\sim Q_i$. Enumerate these as $B_1, \ldots, B_2^n$. Each has probability $1 \div 2^n$; that is, for each i, $Pr(B_i) = 1 \div 2^n$. Then $Pr(\sim Pv \sim B_i) = Pr(\sim P) + Pr(\sim B_i) - Pr(\sim P \& \sim B_i)$, where $Pr(\sim B_i) = 1 - (1 \div 2^n)$ and $Pr(\sim P \& \sim B_i) = Pr(\sim P) \times Pr(\sim B_i)$. Algebraic manipulation gives, for each i, $Pr(\sim Pv \sim B_i) = 1 - (Pr(P) \div 2^n)$.

---

admits of theorizing, or that it is analyzable in nonepistemic terms. Any belief comes out justified once these terms are fulfilled. Ernest Sosa (1985, 1993) seems to identify symmetry with the supervenience of justification on reasons or evidence. Of course, a contextualist who denies that justification supervenes on evidence, because whether or not a given body of evidence is justificatory varies with context, will disagree. This form of contextualism is about justification and is not to be confused with the contextualism about evidence that I defended in Chapter 4, which is consistent with symmetry. I address contextualism about justification in Chapter 7.

Thus $Pr(P) \leq Pr(\sim Pv \sim B_i)$. Suppose that P's (presumptively high) probability justifies P; that is, $J(P)$ (or $JB(P)$, either way). Then $J(\sim Pv \sim B_i)$. As this result holds for each i, CJC gives $J(\Pi(\sim Pv \sim B_i))$. But $\Pi(\sim Pv \sim B_i)$ entails $\sim P$. By CJE, $J(\sim P)$. But if $J(\sim P)$ then $\sim J(P)$; otherwise, by CJC, $J(P\& \sim P)$. Hence the supposition $J(P)$ must be false. Therefore $\sim J(P)$.[8]

It may seem natural that if P is believed and found to be inconsistent with Q, then if $Pr(Q)>Pr(P)$ Q should be believed instead. This principle is unacceptable, for on it nothing should be believed; one should shift one's allegiance away from any belief one ever has. The existence, for any belief, of an inconsistent alternative of higher probability demonstrates that high probability *in itself* cannot be sufficient for justification.

That the alternative the general proof supplies lacks contextual relevance does not affect this conclusion. Possession of uniquely highest probability within a given context cannot be justificatory, because CJC would then justify the contradictory conjunction of inconsistent beliefs justified within different contexts. Certainly the property of being epistemically justified cannot be restricted to a single, privileged context.[9] As probability is not justificatory, the behavior of probability under conjunction does not render epistemically paradoxical the transmission of justification through inference.

There are, however, certain beliefs that we regard as paradigms of justification, even as we believe the basis of their truth to be probabilistic. I believe that an airplane heading into a building will crash, not tunnel through (via the quantum tunneling effect). I believe that water left on high heat will boil, not merely that it is likely to. Surely this credence is not excessive, despite the fact that physical laws governing these processes are indeterministic; it would not strictly violate them for the water to freeze.

I suggest that cases like these depend on a difference between the basis for one's belief and what one (subsequently) learns about the causes of its truth. Beliefs about the ordinary behavior of water are *not* based on laws of statistical mechanics, nor even thermodynamics, let alone quantum mechanics. People who know nothing of any of these theories have justified beliefs about water. They have rough, general beliefs that thermodynamic principles sustain, like the belief that a hot object will warm a cold one. These beliefs, in fact, are *true* (unqualifiedly). It is consistent with fundamental physical laws for there to be exceptions, but there aren't. Indeed, fundamental laws tell us that the probability of there being an exception (in the history of the universe) is vanishingly small. Applying these exceptionless principles to form beliefs in ordinary situations is reliable. A mind that knew nothing about water, that had no true general beliefs about the nature of heat, but started *ab initio* with quantum mechanics and took all doxastic direction from there, would

---

[8] I offer this proof to replace a (faulty) proof given by John Pollock (1983) that depends upon an unjustified (unacknowledged, unsupported) assumption of disjunctive exclusivity.

[9] Contextualists about justification often seem to want to restrict the property of being epistemically *un*justified to a single context, that in which skepticism is in play.

be unrecognizable to us. It is difficult to imagine what such a one would believe, if anything, justifiedly or otherwise.[10]

6.6 Justification and Low Probability

I have argued that a high statistical probability short of 1 for a justified belief neither is the source of the belief's justification nor defeats its justification. Suppose that a justified belief is statistically *im*probable. This would not defeat justification either. For if it did, then CJC would fail. Indefinitely extended conjunctions of statistically highly probable justified beliefs become improbable.[11] For such a conjunction to be justified on my theory, its conjuncts would have to be believed in virtue of the intentional application of reliable methods, not in virtue of any statistical probabilities assignable to them. This requirement will not be met by the presumably improbable belief that no one's car will be stolen, for example. There will be lots of cases in which beliefs to the effect that an individual car will not be stolen are not reliably producible. So it is not clear that this result carries counterintuitive consequences. But beliefs that are justified on my theory must also have a justified conjunction, even if this conjunction is improbable.

Apart from CJC, the low statistical probability of a belief justified under my theory can only result from the application of reliable methods under abnormal conditions. This abuse (so to speak) of reliable methods ought not to impugn them. If there is no reason to believe that a method is being misapplied *now*, then the belief it delivers now can be justified even in the face of a checkered record for the output of the method. Juries rightly trust a witness who turns state's evidence though he has lied before, because his present circumstances, unlike those before, give him no reason to lie and every reason not to. Descartes, in *Meditation I*, rightly rejected the fallibility of sense perception as a general ground for doubt, because the conditions

---

[10] Well, such a one might believe that the moon is not there when no one is looking. It might believe that the pot doesn't boil *unless* you watch it. It might believe that whether or not the universe originated in a big bang depends upon what empirical investigations we now undertake. It might interpret all science instrumentally. It might do all these things at once, and I cannot imagine doing any of them.

[11] I am making a concession here. What kind of *im*probability is this? Is it true statistically that extended conjunctions of statistically probable conjuncts have a low frequency of truth? Not generally. A statistically obtained frequency for the conjunction can be anywhere within the range of possible frequencies compatible with the observed frequencies of the conjuncts. The average value of this range equals the product of the frequencies of the conjuncts (supposing these to be probabilistically independent). It is this average that we identify with the probability of the conjunction, not any statistically measured frequency. The average may not even be obtainable for real, observed sequences, let alone be the frequency actually obtained. In practice, we forget the statistical source of the probabilities of the conjuncts, once these probabilities become available, and simply assume that, as they are probabilities, the axioms of probability apply to them. We assume a common abstract quantity of probability that different sources or measures of probability all agree on. I find this assumption problematic in Chapter 7.

under which sense perception is known to err are circumscribable, as evidenced by further sense perception. A statistically improbable belief can be justified because the source of its improbability lies not in the unreliability of method but in conditions that we have no present reason to fear.

I hope now to have shown that CJC does not make any general skepticism as to the justification of ordinary beliefs the cost of avoiding paradox. Chapter 7 investigates potential skeptical implications of CJE.

Chapter 7
Skepticism

7.1 Discernibility and Evidence

Skepticism is still a potential problem for my theory. Although the dissociation of justification from probability does not vitiate the justification of ordinary beliefs, perhaps skeptical implications of my transmission principles do. If I am justified in believing that I have hands, inference from this belief entitles me to believe justifiedly that I am not the victim of a skeptical scenario in which I do not have hands but merely appear to. But in denying that I am the victim of such a skeptical scenario, I commit myself to the truth of a proposition of whose falsity I could have no indication. There is some plausibility to the principle that one cannot be justified in believing a proposition whose falsity one would be unable to detect. On this principle (DP; "D" for detectable, not for detected), CJE precludes my believing justifiedly that I have hands.

I will not try to prove that I have hands, nor to prove that the belief that I do is justified. Nor will I refute skepticism. I close this chapter with an argument that will satisfy some (including myself) as justification for rejecting skepticism, and I will suggest other anti-skeptical lines of argumentation along the way (primarily in the next section). But I assume that my belief that I have hands is justified, and I think that this is already to assume, in advance of argument, that skepticism is mistaken. For skepticism denies that ordinary beliefs are justified. The justifiedness of one's belief that one has hands is not, for me, a consequence of a theory of justification, but data to theorize about and from. A theory of what justification is does not automatically assume the burden of justifying paradigmatically justified belief, nor of refuting skepticism. After all, in denying that ordinary beliefs are justified the skeptic himself operates with a concept of epistemic justification. It ought to be possible to understand this concept independently of how skepticism is judged.

Rather than the disposition of skepticism, my argumentative burden concerns my transmission principles for justification. In this chapter I am concerned primarily to defend CJE, as in Chapter 6 my primary concern was the defense of CJC. My burden is to show that these principles do not *create* a skeptical problem that is otherwise avoidable with a different theory of justification. Specifically, contrary to both Nozick and Goldman (also Dretske (1971, 2005) and other authors) denying transmission does *not* protect one's (presumptive) justification for believing one has hands from defeat by the possibility that one is victimized in a skeptical scenario. I do not for a moment think that skeptical scenarios *do* defeat ordinary justification, but if they did with transmission they would without it. Accordingly, an interest in avoiding or refuting skepticism with respect to the justification of ordinary beliefs, should one have this interest, is not a reason to resist my theory's commitment to transmission.

According to DP, if $JB(P)$, then if $\sim P$ then $D(\sim P)$. Here the consequent conditional should be read subjunctively, so that detection of the actual falsity of P does not abrogate $JB(P)$; that is, the detectability of P's falsity is compatible with believing P justifiedly: if P is justifiedly believed then were P false, which presumably it isn't, its falsity would be detectable.

DP may recall the requirement of severe testability that I imposed in Chapter 4. There I denied that there could be good evidence for P if there could be no evidence against P. This is a principle of evidence, a constraint on confirmation. DP is a different principle about justified belief; it says nothing directly about evidence. I do not think that my justification for believing that I have hands depends on evidence at all (although in some situations it could); certainly it does not depend on assessing the confirmatory weight of sense impressions. I see that I have hands, I use my hands, and that I do these things is evidence for other things. Evidence has to stop somewhere, and for it not to stop at hands is already skepticism.[1]

Furthermore, the evidence principle—the requirement of severe testability—is in fact satisfied by the belief that I have hands. Someone else's belief that I do (and mine under exceptional circumstances) could be justified by evidence, and there could be evidence to the contrary. The fact that under a skeptical scenario, in which I do not have hands but appear to, there is no evidence available to me that I lack hands, does not imply that there could be no evidence that I lack hands. There could be such contrary evidence available to others, and in nonskeptical scenarios there could be such evidence available to me.

However, under a skeptical scenario, my belief that I have hands does not satisfy DP. I reject DP on the grounds that it is insensitive to how beliefs are formed. Even if perception does not justify rejection of skeptical scenarios, inference from beliefs justified by perception does. DP says not merely that *perception* is unable to justify beliefs of whose falsity there would be no *perceptual* indication. It denies justification by any means to beliefs whose falsity would be undetectable. This is too strong to accommodate justificatory reasoning in general, and scientific reasoning in particular.

First a general point: Paradigmatically justified beliefs carry consequences whose falsity would not affect believing them; one's belief would be unaltered by the falsity of its consequence. My belief that there is an X before me can be jus-

---

[1] Or so it seems to me. In Chapter 5 I noted, with disapproval, that some evidentialists, like Conee and Feldman, want mental states to count as evidence.

tified even though I would still believe this were the object before me to be a nonX indistinguishable, by me under present conditions, from an X (a fake). That the object is an X entails that it is not a fake. Recognizing the entailment, one might yet be reluctant to infer that the object is not a fake. One might demur for reasons discussed in Chapter 5, or just because one is in the grip of DP. But what if one does draw the inference? (What is to stop one, I have asked rhetorically.) Then one has a belief that, though in violation of DP, is justified, for it is reliably formed. If doxastic commitment in this case looks brazen, the moral should be that justification goes to the intrepid (a result that should be congenial to the class of contextualists who suppose that knowledge or justification depends on heedlessness of unrefuted rivals to the truth of one's belief, of which more below).

It is worth pausing to ask whether the same reasoning applies to knowledge. Will a nonskeptical theory of knowledge have the same problem satisfying DP? One is inclined to suppose that this will depend on how the theory treats transmission. But consider the theory of Fred Dretske. Dretske (2005) is for DP and against transmission. But he also holds that conclusive reasons produce knowledge (1971, 2005). Why can't a belief that violates DP nevertheless be held for conclusive reasons, and so be known (or knowable; there may be other requirements that do not matter here)? A conclusive reason for a belief, says Dretske, is a reason one would not have if the belief were false. Suppose that one comes to believe Q by logically (correctly) deducing Q from P, which one knows. Then "$P\&(Q$ is logically deducible from $P)$" is a conclusive reason for Q, since under the condition that one knows that P it is not a reason one could have were Q false. (Dretske does not say exactly what it is to *have* a reason, but it seems safe to assume that if you know a reason you have it.) Nothing prevents P from satisfying DP while Q violates it. Of course, one's theory could simply make DP *itself* a condition for knowledge (formulating DP for knowledge rather than justification), so that DP is automatically satisfied by whatever is known. Depending on how one interprets the semantics for DP's subjunctive condition, one could still get ordinary knowledge. But rejecting transmission is insufficient to guarantee this outcome.

Skeptical scenarios involving systematic deception are just a special case. To those whose inferential alacrity extends to denying fakes but ebbs at the point of systematic deception, I pose the following rhetorical question: If I am justified in believing that I am awake when I am even though I am not justified in believing that I am not awake when I am not, why can't I be justified in believing that I am not victimized in a skeptical scenario when I am not even though I would not be justified in believing that I am victimized if I were?

The point about scientific reasoning is that DP violates scientific method by preventing the evidential discrimination of rival theories that are empirically equivalent—that carry identical consequences for experientially accessible portions of nature. No matter how strong (and severe) the evidence in support of a scientific theory, a kind of skeptical scenario is possible in which the theory is strictly false, but indiscernibly so. For a rival can be algorithmically constructed by fixing all the theory's observational consequences while tinkering with its theoretical mechanisms. Scientific practice rightly accepts the legitimacy of forms of ampliative reasoning that disallow such skeptical constructions.[2]

I also note that DP does not apparently apply to necessary truths, which cannot coherently be supposed false.[3] If DP is restricted to contingent beliefs, does the *cogito* violate it? I am justified in believing that I exist, but the falsity of this belief would not be detectable by me. One might respond that what happens if I fail to exist is not that my belief that I exist is false, but rather that this belief itself fails to exist. If I do not exist, neither do my beliefs. Then DP cannot apply to the *cogito* either. So the *cogito* does not violate DP.

But I think that expelling the *cogito* from DP's range of application is an unnecessarily concessive response. It is the falsity of the propositional content of a justified belief that DP is meant to require to be detectable by the believer. Supposing that I detect my existence, my nonexistence would make a difference in what is detectable by me. DP can be formulated without loss of effectiveness as the requirement that the falsity of the propositional content of a justified belief would make a difference to what the believer can detect. It is the absence of any difference in what is detectable by the believer under skeptical scenarios that challenges the justification of ordinary beliefs.

7.2 Personal Identity and the Mental

I diverge here to question the assumption that DP *does* (would if true) pre-empt justification of the belief that one is not the victim of a skeptical scenario. I do this because I expect that some philosophers will find DP more difficult to dismiss than I do. The possibility of accommodating these philosophers to my position is worth some additional argumentation. I suspect that DP, or something close to it, is responsible for the common conviction that skepticism is deep and compelling. Many epistemologists assume as a condition of adequacy that the skeptic must be given "his due"—that an adequate epistemology must account for skepticism's appeal, for the hold they suppose skepticism to have on the mind. In view of this assumption, it is unrealistic for me to expect the disposition of skepticism I shall propose later in this chapter to be widely persuasive. So while I reject the assumption, it is worth exploring resources available to me even if it is correct. As this inquiry confronts skepticism directly and is not strictly within my argumentative burden, it will be permissible to leave some issues without final resolution.

---

[2] In Leplin (2004), I examine the adjudication of rivalries among empirically indistinguishable theories, and argue that this feature of scientific practice is epistemic, not merely pragmatic. The common consequences of rival theories can differ in their evidential bearing for these theories. So empirical equivalence does not imply evidential underdetermination.

[3] I think that some counterfactuals whose antecedents deny necessities are evaluable, and will mention examples in due course. However, this view is disputatious and nothing turns on it here.

Accordingly, although I prefer to dispel skeptical worries independently, I shall raise some problems for skeptical scenarios under the stipulation that DP is true. Foremost is a problem of personal identity. What makes it the case that the individual who lacks hands, but is deluded into believing he has hands, is *I*? This individual, the victim, has my brain, let us say, or what *was* my brain if I am dead. Surely the survival of the physical organ is insufficient for *my* survival. Surviving organs are transplanted, not to sustain the donor's existence, but to enable others to survive. If the brain is an exception, this is only because of its presumed association with mentality. The victim has to have my memories, dispositions, expectations, preferences, discriminatory responses to the (supposed) environment—enough of these things for continuity of mental life.[4]

Any of these things can change without affecting personal identity; but to the extent that supposed changes in them are abrupt or pervasive, identity comes into question. If *I* am the victim of the mad scientist, must there not remain enough of me to compare for incongruity with the deceptions he implants? The ruse, or manipulation, is supposed to be seamless, so that I notice nothing; yet the impressed impressions, apparent experiences, imagery, and so forth are stipulated to be non-veridical; otherwise I am *not* deceived. I question whether it is fair of the skeptic simply to assume that these two constraints can be met compatibly with sufficient mental continuity to make the victim me. I am inclined to think that victimization satisfies DP after all, because a true skeptical scenario would make me suspicious at least, probably crazy. Will the skeptic claim that no one is ever justified in believing he is not crazy?

If so, we have the makings of a sort of reductio against the skeptic. For, to the extent that a skeptical scenario impairs cognitive faculties, it pre-empts its own persuasiveness. It is unclear how the skeptic can even purport to be mounting an argument that carries the consequence that one is unjustified in believing one is not crazy, if an argument is a coherent line of rational thought that is not crazy. It is unclear how we can take ourselves even to understand the skeptic's reasoning, if that reasoning denies that we are justified in crediting ourselves with a capacity for understanding reasoning. Insanity takes many forms, and this way with the skeptic depends on its taking a form that impairs rational deliberation as much as skeptical scenarios are designed to impair perception. Whether the victim of a skeptical scenario is (necessarily) so impaired seems to me an open question.

---

[4] Doug Long (1992) argues that the skeptical scenario precludes self-reference or self-identification: "Only limited epistemological dislocations occasioned by the sorts of deceptions, dreams, and hallucinations that are possible within the context of our ordinary epistemology are compatible with having beliefs about *oneself*" (p. 72). Perhaps, but Long goes further. He grants that he could come to be no more than a brain at the mercy of scientists able to delude him systematically, but denies that he could have suffered this fate *already*. For having suffered it is incompatible with his *present* ability to identify himself. But then why not dispute the presumption of this ability? It seems to me that Long's position would make more sense if he agreed with me and denied the possibility that it would be he whom the scientists come to delude.

Crispin Wright (1991) argues that he is. According to Wright, the victim cannot follow a line of reasoning; he can only, at best, successively grasp its individual steps.[5] But even granting the distinction, it is unclear why a victim who can do the latter cannot do the former. Still, it seems to me promising to contend that if the skeptic were right about one's inability justifiedly to reject his scenarios, there would be a corresponding inability justifiedly to believe that one has correctly inferred from them that one's justifications of ordinary beliefs are defective. It requires rational reflection to justify believing that a belief whose justification requires rational reflection is unjustified if one is a victim. So if victimization impairs rational reflection, there is no justification for believing that it undermines the justification of such beliefs. For, the belief that it does is one such belief.

I say that this line of argument is "promising", nothing more. It may prove too much. If what prevents me from justifiedly rejecting a skeptical scenario is a general incapacity for correct reasoning, then for all I can justifiedly believe, my ordinary justifications may be defective whether I can infer that they are or not. The skeptic may be unperturbed by his inability to persuade *me* of the defectiveness of my ordinary justifications, if the reason for this inability is that I am unreachable by rational persuasion. He may settle for my inability to defend any claim to justification.

Significantly, however, this defense of the skeptic is not consistently available to epistemologists who credit skepticism with great argumentative cogency. These philosophers will instead have to maintain that (and explain how) one can be systematically deceived by manipulation without being rationally impaired. Otherwise, their very ability to recognize and appreciate the (alleged) persuasiveness of skeptical reasoning can be construed as reason to deny that they are right to find it persuasive.

The problem I raise for personal identity supposes that my mental life has been interrupted and redirected. What if I've *always* been a victim? I am a laboratory experiment none of whose apparent experiences has ever been veridical. Then the question is whether the development of such concepts as veridicality, justification, truth, appearance, identity—even language as such—is possible from a totally solipsistic position. I interpret Wittgenstein in the *Philosophical Investigations* (1958) as showing that the distinction between seeming to reidentify a mental state correctly and actually doing so cannot be drawn from a purely solipsistic perspective. Admittedly, one could be an accurate reidentifier of mental states without the difference between accurate and merely apparent reidentification being manifest to oneself. But the question is whether such unattestable accuracy in one's own case suffices for the acquisition of mental concepts. It is crucial to the concept of a mental state that one recognize the possibility of misidentification, and this recognition depends

---

[5] Wright takes dreaming rather than external manipulation for his paradigmatic skeptical scenario. Although he claims that his argument generalizes, intuitively rationality seems more immediately endangered by dreaming than by the external induction of experiential states. "You're dreaming" suggests that you are being irrational, or at least unreasonable.

on associating the state with behavioral indicators that are strictly insufficient for its correct attribution.

There is also the question whether my (apparent) use of physical-object terms can secure reference, and if so to what. If a causal theory of reference is correct, so that never having been in causal contact with physical objects I cannot use terms that refer to them, then it may be that what I (take myself to) say about them is not false, but either meaningless or true. For, if my terms refer, their referents will have to be things to which I do have access, like induced images or neural inputs from the mad scientist. It then appears that I cannot be (massively) deceived, because the mad scientist's manipulation prevents me from forming the incorrect thoughts in which the deception would consist. Given the semantic facts, the skeptic cannot describe what happens to me in his scenario without speaking falsely.[6]

But am I not deceived, then, about what my words mean, or deceived in thinking that my words have meaning at all? According to Hilary Putnam's externalist view of meaning (1981), the victim of a skeptical scenario cannot actually (and incorrectly) believe himself not to be the victim of such a scenario, for his language cannot refer to such a scenario. What he can believe is only something *about the scenario* (though he does not take it to be about this), to the effect that (within it) he is not victimized, which is true. But even if Putnam is right about this failure of reference, does the victim not still labor under a deception? For, does he not mistakenly *take* his beliefs to be about something that his referential limitations prevent them from being about? Intuitively, what the victim takes his beliefs to be about is not identifiable with the referents of his words.[7] He does not take his beliefs to be about neural inputs or images or any of the things that Putnam will allow his words to refer to. But he does take his beliefs to be about *something*; Putnam does not dispute the victim's possession of beliefs, nor his capacity to believe that he has beliefs. What does he think they are? Evidently, we cannot say without crediting the victim with references that he is supposed to be unable to achieve.

I find this outcome incoherent. I cannot be incapable of referring to what I am capable of believing to be the references of my beliefs. It cannot be a deception that my words have meaning, if I have beliefs whose semantic content they express. On a causal theory of reference, my victimization must consist not so much in deception as in limitations upon my doxastic capacities. Putnam is more permissive than his view of reference really allows. On his view, I cannot even believe that I am not victimized, for I cannot think what I am not victimized by. I am so much the victim that I cannot register my condition. Or, as Timothy Williamson (2000) says, I am so epistemically impoverished that I am unable to recognize my impoverishment. This looks too much like skepticism to credit the causal theory with a refutation of it.

---

[6] David Christensen (1993) extends this point into a general argument for a semantic dissolution of skepticism. His argument does not, however, seem applicable to the case in which my words have secured reference in the (presumptively) normal way and I am then subjected to external manipulation.

[7] And there is precedent for dividing these in the distinction between speaker's reference and semantic reference (Keith Donnellan, 1966).

But there is a complication. If thoughts are mental states and the (presumed) actual and skeptical worlds are mentally indistinguishable, then I am as impoverished in the actual world as in the skeptical. So I cannot understand victimization in the actual world either. This result seems unacceptable. To resist it, we must either deny the mental indistinguishability of the actual and skeptical worlds or deny that belief is (entirely) mental. I incline to the latter position. As the causal theory requires, belief, like knowledge, is partially external in that its semantic content depends on facts independent of mind, which is internal. Then the ability of the scientist to reproduce my mental life is left open.

On Williamson's view, however, knowledge is itself a mental state, and the proper criterion of externality is not mind-independence but physical circumscription—the boundary of the body. Then, if the skeptical scenario robs me of knowledge I otherwise have, it cannot preserve my mental states. If it does not rob me of knowledge, then skepticism is presupposed because I didn't know to begin with. Either way, the machinations of the scientist cannot ground skepticism. This is a crafty maneuver, but I resist the view that knowledge is a mental state.

Williamson claims that knowledge is "prime" in (roughly) the numeric sense— not composed of independent combinable ingredients. He effectively counters attempts to identify a mental component of knowing that falls short of knowing but gets converted into knowing through the addition of something nonmental. But look at knowledge from the outside in. Take a case of knowing and subtract. I, in my office, know where my car is parked. It is in lot 1, (very) far away. Now there is a bomb threat and the police evacuate lot 1, moving my car to lot 2. I no longer know, but my mental states are unaffected. I would appeal to the principle that distant events with which one is causally unconnected (perhaps unconnectable) and of which one can have no awareness do not alter one's mental states. The unknown relocation of my car affects my cognitive states in roughly the way that the unknown death of a loved one affects my emotional states. It is not literally true that I love a person who no longer exists, but this is not a change in me. Certainly, if the mental supervenes on the physical knowledge cannot be a mental state, for moving my car does not change my physical states.

Naturally, if it is established independently that knowing is purely mental then the causal connection principle is mistaken. But the (pure) mentality of knowing is not a consequence of the (presumed) externality of the individuation of concepts, still less of the causal theory of their reference. That the semantic content of belief depends on the external world does not show that moving my car affects my mind. Williamson's argument seems to me to show instead that something is seriously wrong with the supposed category of mentality, and with the way we presume to divide the mental from the physical. If unknown, distant events do change me, something is wrong with the ordinary notion of self-identity, with how we individuate persons and distinguish ourselves from the external world.

As I am not (for present purposes) even committed to the causal theory of reference, let alone to the theory that concepts are individuated by their referents, I am certainly free to deny that knowing is purely mental. Accordingly, I reject Williamson's way with skepticism. The scientist can reproduce my mental life, but in so doing it is unclear that he is manipulating *me*. He can rob me of knowledge, but perhaps does so only by destroying me altogether. To challenge the justification of one's ordinary beliefs, a skeptical scenario may need to challenge the justification of one's belief in one's own existence. If so, then on the supposition that not to exist makes a difference in what is discernible to oneself, we can oppose skepticism with a line of argument that respects DP.

7.3 Relevant Alternatives

Let me return to my principal concern, that my theory not be rejected on the grounds that only by rejecting the transmission principles that it implies is skepticism avoidable.[8] Following long tradition, I represent nonskepticism by the justification of believing that one has hands. Alternative minimum conditions for nonskepticism are (supposing the belief is true) that one be *able* to justify this belief (upon reflection as to the reliability of its source) and that the belief itself be justified. In Chapter 11 I will use the latter condition to avoid requiring that one's justificatory abilities have been exercised. But I am happy with any of them, and I do not think the differences matter to what I want to say in this chapter. It will ease exposition not to trouble over them.

The essential point of all versions is that skepticism denies knowledge or justification. To deny skepticism is to claim knowledge or justification. It is not enough that knowledge or justification be possible, or to claim that they are possible. If justification is possible but we are in no position to say whether it in fact obtains, then we are in no position to deny skepticism. Thus, the thesis that the impossibility of rejecting, or the inability justifiedly to reject skeptical scenarios is compatible with (ordinary) justification does not deny skepticism.

If skepticism is false, then, at the very least, one is justified in believing oneself to have hands. The problem is that this belief entails propositions that (it will be supposed) cannot (in view of DP) be believed justifiedly. Hence, if skepticism is to be false, inference must not transmit justification.

But how exactly does rejecting transmission help to forestall skepticism? How can one be justified in believing one has hands *without* being justified in rejecting skeptical scenarios in which one does not have hands? On my theory, one will be justified in rejecting *any* scenario in which one does not have hands, if one rejects it by inference from one's belief that one does have hands, presuming, as I do, that one

---

[8] Can someone who rejects transmission accept my theory despite its entailment of transmission? Why should the falsity of consequences affect what one accepts, if closure is denied? It is for those who reject closure to figure out the effect on the scope of rational argumentation. Maybe, short of accepting closure as a general principle, one may reason in particular cases that it just wouldn't make sense to suppose that one accepts this without accepting that. But if, as I expect, the supposition turns out regularly to be senseless where this and that are related as per closure, then one might as well face up to the general principle.

believes this justifiedly. I maintain that one fails to be justified in rejecting skeptical scenarios only if one fails to perform the inference.

In opposition, it may be held that one's justification for believing one has hands is independent of what view, if any, one takes of skeptical scenarios, that skeptical scenarios are *irrelevant* to one's justification even though incompatible with the truth of one's belief. According to the opposition (which does not distinguish justified belief from believing justifiedly), the justification of a belief does not require that *all* incompatible alternatives to the truth of the belief be justifiedly rejectable; only certain alternatives that are *relevant* to consider need be rejectable.[9] The relevant alternatives to my having hands are conditions that, (supposedly) unlike skeptical scenarios, would be detectable. They involve my suffering some injury; the ways in which people are *known* to lose hands are those relevant to consider.

An appeal to what is known is, however, suspicious in an argument against the skeptic. And it does not make any essential difference to formulate the position in terms of evidence or reasons short of knowledge. It is equally suspicious to make ways in which we are justified in believing that people lack hands the standard of relevance, if, as I am assuming, skepticism contests the justifiedness of belief.[10] If I am not justified in rejecting skeptical scenarios, then what justifies the assumption that if I lacked hands this would be because of an injury *rather than* because I am the victim of such a scenario? What justifies judgments of relevance for the alternatives to be ruled out? It seems to me that any such judgment *already* rejects skepticism.[11]

---

[9] "All" is italicized in anticipation of the view that universalization in natural language implicates some domain. Nothing outside the domain can serve as an exception. So if I say that all seats are taken, it does not make me wrong that at some other event in some other room of some other building seats are empty. On this view we *can* say that all incompatible alternatives must be rejected. But this is only because not *all* alternatives are included among all the alternatives. As an unrestricted sense of the universal is needed just to state the view, I dismiss it. Of course, *some* uses of "all" implicate a restricted domain, but not all. That is all I need.

[10] One could be a skeptic about knowledge but grant justified belief. Naturally, the skepticism that concerns me threatens justification.

[11] This criticism may be anticipated in Edward Craig's (1989) criticism of Robert Nozick. Craig argues that if skepticism is true then the belief that one has hands does not satisfy Nozick's condition of counterfactual sensitivity to falsity in neighboring worlds, for it is not so sensitive in the actual world, which must qualify as neighboring. He infers that Nozick must presuppose that skepticism is false and so has no argument against it. If I take Craig's point, however, it does not show that Nozick must presuppose the falsity of skepticism as a thesis about knowledge. Rather, he must presuppose that skeptical *scenarios* do not hold. I will argue, further, that the condition that skeptical scenarios do not hold is insufficient to ensure the counterfactual sensitivity to falsity of the belief that one has hands. Anthony Brueckner (1991) thinks it an adequate reply to Craig to point out that on Nozick's analysis of knowledge the transmission principle used in skeptical arguments has no universal guarantee. I dispute Brueckner's reasoning below, but supposing he is right (which supposition is so much the worse for Nozick's analysis, on my view) the issue then is where the failure of a skeptical argument leaves skepticism. Unless there is an argument against skepticism, the failure of an argument for it is insufficient, for an impasse is a skeptical outcome. Nozick might protest that he does not mean to answer the skeptic, but only to protect ordinary knowledge against the irrefutability of skeptical scenarios. But, again, the skeptic's thesis is not that skeptical scenarios are irrefutable; it is that we lack knowledge or justification. As urged

It seems incompatible with skepticism that I should be justified in dismissing as irrelevant certain conditions under which the beliefs I need to come out justified are false. If I am *not* justified in denying that these conditions hold, it is difficult to see what can be my justification for dismissing them.

I grant that there could be a kind of default entitlement to, or presumption in favor of, dismissing them, without such entitlement or presumption constituting or requiring epistemic justification. But then, it seems to me that the status of my believing that I have hands will be similar; it will be presumption rather than justification. It is not clear that such presumption or default entitlement is properly described as "epistemic". How, to recall the discussion of Chapter 5, does the currency of a default entitlement in practice decide its epistemic status? Why isn't default entitlement simply pragmatic, if it fails to constitute epistemic justification? Any good skeptic (Hume, the Descartes of the first *Meditation*) will happily accede to pragmatic entitlements. But even a form of presumption that is (somehow) epistemic rather than (just) pragmatic is not strong enough to counter skepticism. Only if I am epistemically justified in believing that I have hands is skepticism false.

7.4 Proximous Worlds

It is natural to the point of unavoidable to frame this issue in terms of the proximity of possible worlds. Among worlds in which I do not have hands, those in which I do not take myself to have them are alleged to be *closer* to the actual world than those in which this diminution of myself is unnoticeable.[12] In close worlds, I am injured in an accident, undergo an operation, or (a bit farther out) steal something in Saudi Arabia. The criterion of relevance is then proximity. To believe justifiedly, I need only be able to rule out the close worlds in which my belief is false, and this I can do.

The question, then, is what justifies judgments of proximity. If these judgments are measurements from the presumed actual world, there is no argument against the skeptic. The whole point of skepticism is that we are not epistemically entitled to

---

above, skepticism is an epistemological thesis; its disposition must proceed at the second order. The supposed irrefutability of skeptical scenarios is but an argumentative device. Nozick's suspension of transmission disputes only the adequacy of the argument, whereas his claim to knowledge rejects the skeptical thesis itself. The question is what is the basis for rejecting it. I raised this question against Nozick in the introduction, and will return to it below in connection with arguments by Duncan Pritchard and Mark Heller.

[12] Alternatively, worlds in which I do have hands are closer to worlds in which I take myself to have hands than are worlds in which I lack hands. Those who prefer a "safety" condition on justification to the condition that one's belief be counterfactually sensitive to falsity will advocate this formulation. I do not think this matters, because both proximity restrictions ultimately beg the question against the skeptic in the same way. I consider the safety condition in Chapter 8, concluding that the difference in formulations is not epistemically significant.

presumptions as to what world is actual. Of course, in making this point the skeptic does not contemplate default entitlements that do not carry epistemic justification. He means that assumptions as to what world is actual are unjustified. And how are judgments of proximity to be made without such assumptions? Any discrimination among possibilities as to proximity immediately begs the question against the skeptic. After all, if skeptical worries were realized and I were deluded, the closer worlds would be ones in which it continued to appear to me, misleadingly, that I have hands. They would be worlds in which the deception, though varied, is maintained.[13] If I cannot believe justifiedly that I am not deluded, then I cannot justifiedly pronounce as to the relative proximity to mine of worlds in which the falsity of my belief is detectable.

Proximity is normally understood as a relation of similarity. In terms of similarity, the problem is that unless skepticism is false we cannot justifiedly identify the world that worlds we need to rule out are similar to. Even judgments of *relative* similarity generally carry empirical presuppositions that skeptical scenarios can falsify. We are inclined to think of relative similarity as an *a priori*, world-independent relation, so that whether w_i is more like w_j than it is like w_k depends only on what is true in these worlds and not on what is true simpliciter. But suppose that in world w the human body is far more resilient than the human mind, and one is less likely to suffer injury than to hallucinate it. Let w_i be the world we take for actual (in which I am fine). In w_j I am injured. In w_k I hallucinate injury. In w_i, w_j is more similar to w_i than is w_k. In w the comparison among them is opposite. Each *comparison* is correct in the world in which it is made. The comparison correct *simpliciter* is that made in whichever world is actual. Similarity depends on the actual world.[14] Unless skepticism is false we cannot justifiedly say what worlds are similar to ours, nor can we say very much about what worlds are similar to each other. It is difficult to determine how far this restriction goes; I will push it a bit further below.

I grant that ordinary beliefs generate a standard of relevance for alternatives. If my ordinary beliefs are true in the actual world, then similar worlds are worlds in

---

[13] Curiously (and without argument), Nozick assumes the opposite (1981, p. 264): Although if it were false that I have hands I would (nevertheless) not be in a skeptical world, if I were not in a skeptical world I *would* have hands (I would not be in *another* skeptical world). That is, the alternative to the truth of an ordinary belief is a nonskeptical world in which the belief's falsity is detectable. But the alternative to the truth of a skeptical scenario is the world as ordinarily believed, in which I am not deceived. I see no defense for this asymmetry. Proximity is Nozick's proclaimed standard for determining what subjunctive situation would be true were a proposition false. So if proximity determines what would be true if I lacked hands, it must also determine what would be true if a given skeptical scenario were false. Instead, Nozick switches his standard from proximity to remoteness. If, as Nozick assumes, skeptical scenarios are the remotest alternatives to the truth of ordinary beliefs, so that they would not hold if ordinary beliefs were false, then the closest alternatives *to them* are also remote.

[14] If this conclusion is correct, then Brueckner's (1991) defense of Nozick fails. For some justified presupposition as to what world is actual is necessary to construct a counterexample to transmission.

which they remain true, for the most part. Worlds that falsify them systematically are radically dissimilar, and the fact that in them I believe incorrectly may be ruled irrelevant to my justification. If it is only in similar worlds that the justifiedness of my beliefs requires them to be counterfactually sensitive to falsity, then that they are not counterfactually sensitive to falsity in skeptical worlds does not matter. These are not among the similar worlds. Were I *in* a skeptical world my beliefs would not be justified, for *then* the similar worlds would include ones in which the counterfactual standard is violated. But so long as I am not in such a world, so long as my ordinary beliefs are true, they are justified and *so is my belief that I am not in a skeptical world*. For the justification of this belief is to be reckoned against the same similarity class as ordinary beliefs, and within this class the belief that I am not in a skeptical world remains true. There is no similar world for the belief that I am not deceived to be false in, so it cannot matter that I would believe incorrectly were I in such a world.

Duncan Pritchard (2002) claims, in effect, that this result establishes closure for knowledge under entailment. The qualification is that as Pritchard is concerned with knowledge rather than justification, he requires not just counterfactual sensitivity to falsity but the stronger condition of tracking; counterfactual sensitivity to truth is added.[15] Although tracking is not in general closed under entailment, entailment does preserve tracking, thinks Pritchard, across a class of similar worlds. The closure I require for justification does not follow, because the condition that fixes the similarity class is that one's ordinary beliefs be true. Knowledge delivers this condition; justification does not.

The more pertinent problem, however, is that Pritchard's argument does nothing to oppose skepticism, for just the reason I have been advancing. It does nothing to show that one does in fact know that one has hands. It shows at most that one would (be in a position to) know that skepticism is false if one *did* have such ordinary knowledge. And it shows this only on the (dubious, to my mind) assumption that none but similar worlds need ruling out. But ordinary knowledge is what the skeptic disputes.[16]

Admittedly, if Pritchard is right the skeptic cannot show that we lack ordinary knowledge either. For knowing, as Pritchard (externalistically) understands it, is compatible with the indiscernibility of the skeptical and presumptively actual worlds. But if the matter is a standoff, the skeptic seems to have the better of it. We *might* know that the skeptic is wrong, but there's no way to tell. To *claim* that the skeptic is wrong, it is not enough to claim to have hands; we must claim epistemic entitlement to claim this. We must claim to know (or, as I have it, to be justified in believing) this. The skeptic's challenge to us is to justify this latter claim. Again,

---

[15] Chapter 8 argues that neither knowledge nor justification requires tracking.

[16] That Pritchard's concern is knowledge rather than justification makes no essential difference to my complaint. A reformulation of Pritchard's argument in terms of justification fails against skepticism about justification for the same reason that the original argument fails against skepticism about knowledge. This is to be expected if skepticism about justification is the stronger position.

skepticism is not a thesis about hands; it is a thesis about knowledge or justification. An adequate answer to skepticism must operate at the second order of epistemic properties.

In fact, however, the matter is not a standoff; I think that the skeptic has a further argument against Pritchard. It is the truth of one's ordinary beliefs that is supposed to establish the similarity class of worlds. Then one knows if what one believes in these worlds matches (tracks) the truth; in particular, in these worlds one either believes correctly or not at all. It is open to the skeptic to argue that for similarity classes thus established, one *cannot* know.

For the condition that an ordinary belief be true does not suffice to make the actual world dissimilar to a skeptical world. I could have hands in a world in which covert brain abductions are commonplace. My brain happens not (yet) to have been abducted, but I do not know that. I do not know that I have hands, nor does Pritchard's account credit me with knowing. There are similar worlds in which I don't have hands but believe that I do, assuming that I believe this in fact. To expel skeptical worlds from the similarity class, Pritchard needs not just the truth of an ordinary belief, but also the systematic veracity of one's belief system. If one cannot know just a little, if it's got to be a lot or nothing, the onus of argument is all the more in favor of the skeptic, who may then pronounce Pritchard's closure result vacuous for want of satisfied antecedents.

Mark Heller (1989) prefers realism to similarity as a standard for the proximity of worlds. Heller thinks that the justifiedness of one's belief requires only that one not hold it in any *realistic* world in which it is false. The skeptic errs in including unrealistic worlds among those that one must be justified in ruling out. My criticism of Pritchard applies to Heller. What is realistic depends on what is real. In a skeptical scenario the worlds the skeptic includes are not unrealistic, so we cannot dismiss the skeptical scenario Heller's way without presupposing what the skeptic disputes.

In (1999a,b) Heller agrees that judgments of proximity of worlds presuppose what the skeptic disputes. His point is that so long as we are not *in fact* in a skeptical world, we have lots of knowledge by ordinary standards for knowing. So the skeptic cannot show that we do not know. But neither can we show that we do. Heller's argument leaves unchallenged the thesis that, by ordinary standards for knowing, standards (supposedly) requiring our beliefs to be counterfactually sensitive to falsity only in proximous worlds, we cannot know that we know. For we do not know that the world that proximous worlds are proximous to is not a skeptical world. As I said about Pritchard, this sounds like skepticism to me. It does not *refute* the skeptic to show that he could be wrong.

Furthermore, Heller's concession that judgments of proximity presuppose what the skeptic disputes is unconvincing. He claims that skepticism sets standards too high to be of interest, so inclusive of possibilities that the differences of epistemic condition that matter to us get lost. When I need my car, the possibility that it has been towed matters, not the possibility that I have been deceived into believing that I own one. But if we *are* in a skeptical world, then a standard that includes skeptical worlds need not be high or inclusive. Does Heller concede that attributions of knowledge, analyzed his contextualist way as the goodness of one's epistemic condition, presuppose what the skeptic disputes? If so, then why does he think that restrictions on proximity answer skepticism? If the actual world is a skeptical world, then whatever the restrictions on proximity skepticism is true.

As a point of exegesis, Heller is not trying to *refute* skepticism. Refutation is not what Heller thinks that answering skepticism means. He does not contend that the skeptic's excessive standards are inadmissible; they violate not facts but conventions, the conventions of the relevant linguistic community (1989). The skeptic simply operates with a different conception of epistemic entitlements than the rest of us. So he is not so much wrong as off the subject. Does this position fare better?

I can only suppose that Heller means to be referring to his *own* linguistic conventions (if this is possible), since under a skeptical scenario no linguistic community is available for him to appeal to. But by anyone's standards, from a skeptical position possibilities of delusion are not unrealistic, overly inclusive, or set too high to affect distinctions of epistemic position that matter. It is quite true that refuting the skeptic is not an appropriate step in a recipe for chocolate cake. Someone whose objection to proceeding to fold the beaten egg whites into the ground hazelnuts is that we have not yet justified the belief that these ingredients exist is changing the subject. But Heller's subject *is* skepticism, not cookery. Moreover, a context in which raising skeptical possibilities *does* change the subject may nevertheless require justified rejection of them. It does not answer a question to point out that it differs from the question under discussion.

One might try to finesse question-begging assumptions as to what world is actual by taking the world with respect to which proximity is measured to be one's natural doxastic world. This is the world specified to the point of one's natural beliefs, instead of the world that actually realizes one's beliefs. However, the natural doxastic "world" is in fact not a world at all but a class of worlds that includes skeptical worlds that leave one's natural beliefs unaffected. The skeptic's claim is that one cannot justifiedly discriminate among the members of this class. It simply begs the question to assume that the only worlds I need rule out to justify my beliefs are worlds outside this class.

I have one further point to make about reliance upon assumptions as to what world is actual. Even availing ourselves of such assumptions—notwithstanding my complaint that to suppose them justifiable begs the question against the skeptic— may not suffice to disqualify skeptical worlds from the domain of the proximous. The further problem is that the notion of a skeptical world is not all that clear. It is a necessary condition of my not being in one that I have hands, but is this condition sufficient? Surely not, but then what is?

Consider the world in which there are hands, apples, zebras, tables and chairs, just as there appear to be, but aliens monitor our reasoning and should we approach a correct understanding of knowledge or justification, should we make any real progress in epistemology, they will send us back to square one (a world in which a book like this can't get published). This is a kind of skeptical world, in that within it skepticism is irrefutable. Of course, one could include the nonexistence of the aliens among our beliefs about what is actual. But then we pick something else. Maybe a computer virus projects hallucinogenic imagery upon the screen of any computer into which a truth of epistemology is recorded. Relative to any circumscription of what we believe, there will be something to pick. And the standards for proximity we have been presented with all require some such circumscription.

7.5 The Status of Subjunctive Conditions

My objection to assumptions as to what world is actual is the more acute if, as I recommend, one takes talk of possible worlds to be metaphorical for talk of counterfactual situations. Possible worlds do not, in my view, explicate counterfactuals, because their identification is parasitic on references to the actual world. We can say what possible world we are talking about only to the extent of including it within the class of worlds in which a variation of some feature of the actual world is stipulated. It is *this feature* that differs in the worlds in question, not something else that resembles it to the point of the variation. To take an example from Saul Kripke, to whose conception of possible worlds I take myself to be deferring, in saying that Humphrey could have won the election I am speaking of Humphrey *himself*, the actual man, not of someone who *did* win an otherwise similar election in a possible world otherwise similar to the actual, and who otherwise resembles Humphrey.[17] As I understand possible worlds, they do not supply a categorical ground for subjunctive conditions. If possible worlds are derivative, as I contend, then any comparisons among them as to proximity, whether to the actual world or simply to one another, presuppose the perspective of the actual world.[18] But the justifiedness of assuming this perspective, the justifiedness, that is, of assumptions as to what is actual, is just what the skeptic challenges.

Is it legitimate to use subjunctive conditions in one's theory without supplying them with a categorical ground? Some philosophers believe not, and several lines

---

[17] If possible worlds are ontologically primitive, can the same individual exist in more than one of them? Did Humphrey himself win the election in another world? Maybe there is some basis for this idea in the many-worlds interpretation of quantum mechanics, but I cannot make sense of it. Which future does an individual with different futures predict? Probability is no help if the actual individual occupies both probable and improbable futures. I shall not pursue this idea here.

[18] It might be thought that some proximity comparisons among possible worlds are nonempirical, and so are independent of judgments as to what is actual. A world in which Humphrey wins by just a few votes is closer to a world in which he wins by just a few more than to one in which he wins big. But numerical proximities are not proximities among worlds. What if the world were such that the easiest or most effective way to shift votes does so only in large blocks? What if it were harder to get Humphrey's margin up only slightly than hugely? Maybe voters follow a kind of herd instinct, and are open to influence only indirectly, via their identification with the herd. We assume an enormous amount about the way things actually work in comparing possibilities; the comparisons are not *a priori*. I fault Heller's (1999a) on this point. Of course, Heller does not say that the ordering of possible worlds is *a priori*; he takes it to be contextual. But he takes assumptions about what world is actual to affect only the center of the orbits of worlds, not their order or distances. Thus he assumes that proximity relations can be decided *a priori*.

of criticism have been proposed.[19] I believe so, on two counts. First, subjunctives are necessary to understand nomological necessity; nomological necessity requires resilience under subjunctive conditions. And the concept of nomological necessity is needed to state fundamental theories in the sciences. This is most evident for theories that propose second order laws about laws. For example, special relativity constrains whatever the laws of nature turn out to be; the only law it states is the independence of the velocity of light. And it is a law of general relativity that the equivalence principle is a law, and not the mere empirical regularity it is taken for within Newtonian theory. Symmetry principles in contemporary physics are constraints on laws; they are laws about the mathematical properties of laws. We have, as yet, no satisfactory way to distinguish laws from regularities that does not depend on the fact that laws support counterfactuals. And we have, as yet, no satisfactory analysis of counterfactuals in purely categorical terms. This does not prevent science from using the concept of law. Why must a philosophical theorist wait when scientific theorists do not?

Second, I think it is time to take seriously the possibility that subjunctive properties are ontologically on a par with categorical properties. Along with abstract mathematical properties, propensities, dispositions, chances, and causes, subjunctive properties are needed in theorizing. This is not always evident, because the subjunctive property of an actual entity is sometimes expressed in the indicative mood in terms of properties or relations of hypothetical entities. For example, causation within a population is subjunctive: the effect would have a greater frequency if all of the population were exposed to the cause than if none of the population were exposed to the cause. But it is common to state this indicatively by comparing the frequency of the effect in hypothetical (nonexistent) experimental and control populations. This is essentially the semantics of possible worlds, which are described in the indicative mood, and which I have argued are derivative. The fundamental property is subjunctive. I propose to regard subjunctive properties as theoretical

---

[19] Robert Fogelin (1994, pp. 66-75), identifies cases in which subjunctives are indeterminate in truth-value, and he identifies difficulties in specifying truth conditions for them. I agree with Fogelin on these points, but will argue that they do not support a prohibition on the use of subjunctives in philosophical analyses. Colin McGinn (1984) denies that counterfactuals are primitive; they are true only in virtue of categorical propositions. But what (little) McGinn says in defense of this thesis is consistent with a reciprocity of dependence, such that with respect to categorical propositions some counterfactuals are derivative and some are primitive. Nothing McGinn produces by way of argument precludes there being categoricals whose truth-value depends on counterfactuals. It could be that whether the philosophical analysis of a concept legitimately proceeds counterfactually depends on what concept one is analyzing. Robert Shope (1978) shows that theories of subjunctive conditional form can come to grief from complications that follow upon the supposition that their antecedent conditions are realized. An example (of my own) to convey Shope's idea is this analysis of the adequacy of evidence in terms of justification: evidence e for P is adequate $\equiv$ if one believed P on the basis of e, one's belief would be justified. For $P =$"I believe nothing on the basis of evidence", the analysis implies that no e can be adequate for P. The analysis fails because the supposition that one believes P on the basis of e falsifies P even if e is adequate for P. I consider in Chapter 10 how such complications affect my theory. Here I observe that Shope's point does not impugn reliance on subjunctives in general, nor does Shope claim that it does.

properties, the disposition of whose ontological status depends on the fortunes of theory and is not to be settled *a priori* (except in so far as the fortunes of theory may be *a priori*).

Furthermore, it is not clear that a subjunctive analysis of reliability without a categorical ground is any worse off than a probabilistic analysis of reliability. For probabilities, too, are theoretical entities. There is no adequate reduction of them to anything more fundamental. Different interpretations of probability are used for different purposes, just as physics uses incompatible models of a single theoretical system in solving different problems. No single interpretation of probability is applicable universally. This is why Goldman needs both frequencies and propensities for his theory of justification. Worse, different interpretations typically support conflicting probability assignments in the same situation.

Consider, for example, the probability that the statistically obtained frequency for the truth of a conjunction, the proportion of cases in which the conjunction is *found* to be true, equals the product of the probabilities, obtained statistically or otherwise, of the conjuncts, supposing these to be probabilistically independent. This probability cannot be 1. It is generally quite small, decreasing with the length of the sequence of cases over which observations are made. The frequency with which a coin (in fact) yields heads differs from its propensity to yield heads. It is, in fact, *improbable* for distinct interpretations to agree in their assignments of probability. And *this* (im)probability is conceptually independent of the choice of interpretation, for it is a result upon which all interpretations agree; they agree to disagree.

That we lack a general theory of the truth-conditions for subjunctive conditionals is insufficient reason to ban them from theorizing. We have no satisfactory general theory of the truth conditions for theorems of mathematics, for the assignment of probabilities, for ascribing propensities, or dispositions, or nomic necessity. We do not even have (nondisputatious) truth-conditions for ordinary, nonmaterial indicative conditionals. Despite difficulties of application, all of these concepts are effective theoretical tools. They are understood pretty well, despite the inability in some particular cases to determine exactly how they apply.

If the truth-value of the subjunctive conditional of my theory is elusive in a particular case, then my theory will not decide justifiedness in that case. But a theory of what justification is does not have to decide whether particular beliefs are justified, or whether they are held justifiedly. The theory may tell us correctly that these verdicts depend on further information that is unavailable. The application of my theory does not require us to know exactly or fully what would have happened in a case of belief-formation were the belief formed to have been false. The case need only be described to the point of determining whether the belief would still have been formed by the method intentionally used. Satisfaction of my subjunctive condition leaves further detail indeterminate. But the detail required may already be fairly rich—richer than we often have, possibly even too rich for full articulation. In some cases we may be unable to decide whether a belief is justified, and also unable to say definitively what further categorical information would enable us to decide.

Such limitations are characteristic of theorizing. We operate with probabilities without assigning specific probability values, nor even being able to specify what categorical information would determine specific values definitively. We use the concept of necessary truth, although necessity has numerous modes among which the distinctions are rough and disputatious. In ethical theory we explicate and use concepts of right and wrong, despite our inability to say in particular cases exactly what factual information would suffice to decide unequivocally the morality of an action. For example, the tenability of a consequentialist ethical theory is not thought to require determination of the consequences of an action, nor even determination of what it takes for something to be the consequence of an action, let alone what it takes for a consequence to qualify as good or bad. I do not see that theoretical reliance on subjunctive conditions poses problems significantly different from those in other such examples. Let us not enforce selectively standards that enforced uniformly would foreclose theorizing altogether.

7.6 Standard Contextualism

Is there an alternative to epistemic perspective as a standard of relevance for possibilities that falsify a belief that the justification of the belief requires ruling out? Remembering that justification pre-analytically involves rationalization, one would expect it to have a strongly pragmatic dimension. Whether I am justified in leaving the child depends on what is at stake. Am I going out for a pint, or to save the world? Jurors are held to a higher standard of evidence than spectators, because theirs are the beliefs that decide the defendant's fate. Perhaps I am justified in believing that O. J. Simpson is guilty, but they are not. Our contexts are different, and standards for justification are contextual. Maybe context determines what are the relevant possibilities that the justification of a belief must rule out. In particular, maybe the contextual irrelevance of skeptical possibilities to ordinary belief protects ordinary justification against skepticism.

Here it seems to me that the analogy of belief to action breaks down. The justification of action is certainly contextual, and what varies in these contexts is not the justification of one's belief but the justification of acting on one's belief. The finding of guilt is an action that the jury takes. Because of what is at stake, jurors must not take this action, even if they believe justifiedly, just as I do, that the defendant is guilty, unless the appropriate evidential standard of reasonable certainty is met. Indeed, their charge is not to judge guilt or innocence at all, but only to judge the adequacy of the state's case. Even a juror not convinced of guilt is charged with delivering this verdict if the basis of his doubt is philosophical only. Philosophical doubt is unreasonable.

But is there not a contextuality to justification in ordinary language? The road is flat if you're driving an Abrams tank, but not if you're riding a tricycle. The refrigerator is empty if you're looking for food, but not if you're verifying the installation of shelving. There are two claims to consider. One is that justification is variable in the way of flatness and emptiness. My belief is justified if the matter is but one of idle curiosity, but not if I am responsible to others for its truth. The second claim is that whether or not justification *as such* exhibits variability, it acquires variability in application to concepts that have it. We should grant that a *finding* of flatness, or emptiness, is justified in one context but not another.

My inclination here is to contest the data. I read ascriptions of rough, qualitative descriptions, like "flat" or "round", differently from ascriptions of justification. Qualitative descriptions admit of different standards, and what varies with context is the appropriateness of the standard. The choice of standard is interest-relative. This much is true of justification. A belief can be more or less justified, and we rightly require greater justification where more is at stake. But the interest-relativity of standards does not make the property attributed contextual. Contextuality requires possession of the property to depend upon some changeable, extraneous feature of the conditions of attribution. "Expensive" I think is contextual. Fresh fish expensive at the coast is, at the same price, a bargain inland. Unlike expensiveness, it is not the property of being flat or round whose possession changes with context, but the degree of it that matters. The road is flat enough for one purpose, but not another. "Empty" I think is simply ambiguous.

The extent to which a belief is justified can certainly matter. But a belief sufficiently justified in one context while insufficiently justified by the higher standard of a more demanding context is not, on that account, *un*justified in the latter context. Relocating me to a more demanding context does not affect my epistemic entitlement to hold the belief. Learning how much (more) is at stake, I change my behavior, not my mind. I become cautious, reluctant to authorize action on the basis of my belief. I do not become unconvinced. My doxastic attitude is stable across contextual changes in standards.

This phenomenon, I submit, is contrary to what one would expect were the property of being epistemically justified contextual, whether intrinsically or in application to contextually variable descriptions. One would expect the loss of justification pursuant to the shift to a more demanding context to affect conviction. Of course, one could lose justification without realizing this. And one could (I will for the sake or argument suppose) realize that one's belief has lost justification but persist in the belief despite this. So the stability of doxastic attitude to which I call attention is not decisive. But I claim that this stability is evidence that justification is not lost. It is at least *unnatural* to maintain full conviction, to entertain no doubt or misgiving, while recognizing that it would be a *mistake*, in one's present, altered circumstances, to regard one's belief as justified. And doubt, misgivings, loss of conviction, are incompatible with belief, as I understand it. Doxastic stability strikes me as an important unrecognized problem for the thesis that justification is contextual.[20]

---

[20] Chapter 5 called attention to a natural variation in credulousness. Is there, in addition to this variation, a contextuality to standards for investing credence, such that the inclination to believe responds to changes that preserve evidence and reasons? Could it be this contextuality that the contextualist mistakes for a contextuality in the epistemic status of what is believed? I think this is a possibility worth recognition and pursuit, but the doxastic stability to which I call attention

A comparison with the effect of contextual change on knowledge attributions is pertinent, especially if knowledge requires justification. Never mind, now, whether knowledge is analyzable in terms of justification and other conditions; just suppose that if one knows that P then one (possibly *in virtue* of knowing) is justified in believing that P. Consider a case in which it is natural and reasonable to attribute knowledge, but then something happens that invests P with greater moment. As a consequence, it becomes unreasonable, or premature, to attribute knowledge without further inquiry.

Let the original case be the diagnosis, by a general practitioner, of my chest palpitations as indigestion. This satisfies me, until I learn of a new treatment for heart disease that works wonders, but only if administered immediately upon the onset of symptoms. The stakes are raised, for if I demur and the practitioner turns out to be wrong, it will be too late to avail myself of the new cure. I consult a specialist. Of course, if the specialist determines that the palpitations are due to heart disease, then the general practitioner did not know. I was mistaken in crediting him with knowledge of the cause of the palpitations. But what do we say if the specialist concurs?

We can reject the possibility that although the general practitioner was right, he did not know. If it takes a specialist to know, then any basis for belief that is improvable is inadequate for knowledge. It would be incorrect to attribute knowledge to the specialist, if someone is more expert still. As this result is unacceptable, one might want to say that it is correct to credit the practitioner with knowledge in the original context, but not in the more demanding context of urgency created by the availability of the new treatment. I wish to reject this option as well. It misses the crucial point that the specialist's findings *vindicate* the judgment of the general practitioner. Upon learning of the new treatment, it becomes reasonable for me to withdraw my attribution of knowledge to the general practitioner pending further inquiry. But this withdrawal is provisional, the attribution to be reinstated if further inquiry confirms it. What one should say is that the general practitioner knew all along, and that's what is learned from the specialist. There is no need to posit any context in which the general practitioner is ignorant.

The addition of information that raises the stakes is incidental. We get the same result if, with no further information, the practitioner's diagnosis satisfies me but not you. You point out that my symptoms could be serious. You know of cases in which symptoms like mine were taken for indigestion but turned out to be a heart condition requiring immediate treatment. Alarmed, I agree to consult a specialist. If the specialist concurs, I was *right* to accept the practitioner's judgment; he knew after all, and you were wrong to doubt him. I do not mean that your doubt was inappropriate; I mean that it was mistaken.

---

is a problem for it. An increasing reluctance to form beliefs where more is at stake suggests an increasing inclination to suspend beliefs once formed.

Now, I submit that justification fares similarly, as one would expect it to if justification is required for (or by) knowledge.[21] A difference in the demands of context does not affect justification. The more demanding context makes it appropriate, not to deny justification, but to seek confirmation of what is justified already. The contrast with knowledge is that it is *not* reasonable (it makes no sense) to ask for confirmation while attributing knowledge. So we suspend the attribution of knowledge; we place it on temporary leave, but leave *with pay* (so to speak), with a continuing presumption of correctness but inoperative pending inquiry. It *is* appropriate to require confirmation of a belief that we continue to credit with justification. All that contextual change suspends in the case of justification is the pragmatic reasonableness of acting on a belief that nevertheless is (remains) justified.[22]

This stability of doxastic attitude shows that not only are the properties of being justified and knowing independent of context, but so are the meanings of justification and knowledge. It is not the truth-conditions for knowledge or justification that change with context. In the case of justification, what changes is willingness to act on justified belief, since this willingness varies with the level of justification that context demands. In the case of knowledge, willingness to attribute knowledge changes along with willingness to act. But a change in willingness to attribute knowledge is not a change in the truth-conditions for the attribution. If I insisted (in resistance to your entreaties) that the general practitioner knows what he is doing and there is no need for a specialist, I would (*ex hypothesi*) be right. If truth-conditions for sentences attributing justification or knowledge shifted with context, then one would expect claims to doxastic entitlement to be withdrawn, and belief correspondingly suspended, with the shift to a more demanding context. The change one finds is in pragmatic entitlement only.

This argument against contextualism from doxastic stability applies both to the justification of belief and to the justification of believing. The shift to a context that raises the stakes for being right about P does not suspend either form of justification with respect to P. For it does not affect the reliability of the process by which P was formed. And it does not impugn one's reason for believing this process reliable.

---

[21] One would not expect this if one thought that being justified means reaching a context-sensitive threshold in the strength or degree of one's justification, but that knowledge requires, invariantly, the strongest possible form of justification. Then contextualism about justification is compatible with invariantism about knowledge. I pass over this option, because on my view comparisons of degree of justifiedness presuppose justification *simpliciter*.

[22] I would apply this result to contextual contrasts invented by other philosophers, like Stewart Cohen (1999) and Jeremy Fantl and Matthew McGrath (2002), except that in their examples I do not share their intuition that the less demanding context supplies epistemic justification. Therefore, if evidence is the same in the more demanding context, as it is supposed to be in the Fantl-McGrath cases, I dispute the intuitiveness of justification being present at all. What I do maintain is that *if* justification is present in the less demanding context, then not only one's belief in that context but also its justification in that context are sustained by conducting the additional, presumptively confirmatory, scrutiny that contextual change demands.

It may induce one to seek a better reason. But this is to seek further or stronger justification, not to create justification where it is (has become) absent.

Contextual shift may also induce one to verify the normalcy of conditions. Before it was enough that there were no grounds to suspect abnormalcy; now one wants to confirm the presuppositions of one's method. Successful scrutiny of these types improves the degree of justification; it does not affect its possession. And a belief that fails such scrutiny will be suspended, so that its justification is not then at issue. That inquiry pursuant to a change of context can change what is believed does not support the contextuality of justification.

If further inquiry determines that one's belief was *not* reliably formed despite one's good reason for having believed it reliably formed, then it was not justified despite having been justifiedly believed. Thus, the shift to a more demanding context can lead to change in what is justifiedly believed, without changing the justificatory status of the belief itself. That is, having determined one's belief to be unreliable, one might no longer believe what one believed justifiedly. But it is not contextual change that makes this difference; rather, contextual change motivates inquiry whose possible outcome makes this difference. What contextual change itself affects remains only the degree or extent of justifiedness requisite to action.

7.7 Variations on Contextualism

There is an elegant formulation of the contextualist position that does not require variation in the truth-conditions for attributions of epistemic properties. Truth-conditions accompany meaning, and the meanings of epistemic terms are stable across context; what changes is the proposition expressed. This formulation normally applies to knowledge, but we can try it out on justification. We should do so respecting my theory's division of justified belief from justified believing. Instances of '*S*'s belief that *P* is justified' and '*S* is justified in believing *P*' then express new propositions in a new context that raises the stakes for being right about *P*, or, for generality, otherwise promotes the salience of heretofore neglected rivals to *P*. These epistemic attributions are indexical in the manner of "I am sad (now)", which expresses different propositions with changes of speaker and time without change in the truth-conditions for being sad nor in the concept of sadness.

My principal reaction to this contextualist position is that it appears entirely *ad hoc*. I see no independent indication of such indexicality in the attribution of epistemic properties, no reason to posit such indexicality independent of the anti-skeptical program of protecting justification in ordinary contexts against its loss in skeptical contexts. I further question what is gained by abstracting truth-conditions away from the proposition a sentence expresses on a particular occasion of its use. The distinction between the contextuality of truth-conditions and the contextuality of propositions seems to me arbitrary. Of course, the meaning of "sad" is not contextual, but why isn't the contextuality of the referent of "I" enough to change the truth-conditions for "I am sad"? Why not say that the truth-conditions for "I am sad" depend on the referent of "I", and, for that matter, the time of utterance? The difference that the introduction of indexicality makes to the contextualist position seems to me inessential.

If this criticism is correct, then my objections to claiming contextuality for the property of being justified, to the thesis that a belief justified in one context is unjustified in a more demanding context, continue to apply. On the present version of contextualism, no single attribution of justifiedness varies in truth-value. Rather, the attribution of justifiedness is itself contextual. But the doxastic stability to which I have appealed counts equally against this view. If my belief is not justified in the more demanding context, then why do I sustain it? The answer must not be that I fail to recognize in the more demanding context that my belief is unjustified. There is no reason for a change in demands to change the recognizability of the property of justifiedness.

Put the matter in terms of my justifiedness rather than that of my belief. If it is true that I believe P justifiedly, and I continue to believe P despite learning that more is at stake in being right about P, then the additional burden on being right does not seem properly interpreted as it now being false that I believe P justifiedly. That the present falsehood is consistent with the former truth because different propositions are at issue does not diminish the implausibility that the contextual shift leaves belief unaffected.

Can contextualism be protected against objections based on doxastic stability? Contextualism may be construed as a semantic or linguistic thesis, rather than as an epistemological thesis; as a thesis, say, about "knowledge" or "justification" rather than knowledge or justification. It is conditions for the use of "knowledge", for the application of the term, that change with context, and changes in these conditions do not imply changes in what one believes.[23] Contextualists are simply describing how the salience of ever more recherché alternatives to the truth of what one believes affects linguistic behavior.

I offer three thoughts about this linguistic turn. First, if contextualism is *not* an epistemological thesis, then I do not see how my argument can be affected by it. In particular, my theory of epistemic justification is not and, for reasons advanced in Chapters 1 and 2, should not be a theory of how the word "justified" is used. But, second, I do not see how contextualism could *fail* to be epistemological. Presumably, "knowledge" refers to knowledge, and "justification" to justification. A thesis as to the conditions for the application of the terms must carry consequences for the ascription of the properties. If it is correct in one context but not in another to describe the subject's condition by using the term "knows", then in the one context but not the other the subject knows.

So transparent is this observation that I would accuse the semantic move of evasion, if it were not more reasonable to suspect myself of oversimplification. Perhaps I have neglected the complication that "knowledge" refers to lots of different things. Of course, all these things are knowledge in virtue of being referents of

---

[23] See, e.g., Keith DeRose (1995, 2002).

"knowledge", but in other important respects (which, exactly?) they are dissimilar. So when one uses "knows" to describe the subject in one context but not in another context, one does not withhold from the subject in the latter what one attributes in the former. Something else is withheld, for in the latter context "knows" has a different referent. Concerned about skeptical possibilities, one does not say "knows" of the subject; reengaged by quotidian concerns, one says "knows". As a change in circumstance presses possibilities ordinarily too remote for concern, the usage of "knows" shifts.

However, if a difference in one's concerns changes what "knows" attributes, it is unclear why the usage of "knows" should vary. Why not continue to say "knows" of the subject, meaning by it whichever of the differing attributions made by "knows" fits the context? Yet the alleged linguistic phenomena cited to support contextualism are changes of inclination to use the word. If the use of the word and the proposition the use of word expresses both vary with context, determining the net effect across contexts looks tricky. How do we know that the effects do not cancel? Contextualists might be well advised to settle for just one of these variables.

Keith DeRose (2002) exemplifies the first choice. He argues that the truth conditions for knowing change, and one's willingness to use "knows" changes as a result. For example, he expects everyone to grant him knowledge, on ordinary evidence, that the bank is open Saturday when it doesn't much matter, but to deny him this knowledge on the same evidence when it does much matter. He thinks he can refute the invariantist's interpretation of such variation as a contextuality of appropriate assertability, rather than of the truth conditions for knowing.

I propose a different interpretation not subject to his refutation: Lots of knowledge attributions are understandably loose. They can be mistaken in ways that would be readily recognizable to the attributer, upon reflection and inquiry that it is usually inappropriate to undertake. Suppose I grant that DeRose knows when little is at stake. I then learn that the bank recently cancelled Saturday hours and only just today (Friday) rescinded this policy without yet announcing the restitution of Saturday hours. As this additional information changes neither the stakes for being right nor (we may consistently suppose) the likelihood of error (they could have changed their minds again by now), there is no reason to assess a change of context. Yet with the additional information, upon reflection, I would retract my attribution of knowledge; I do not think DeRose knows in this case.[24] Conversely, an initial denial of knowledge when much is at stake may be retracted as an overreaction to risk. Upon reflection, it becomes natural to decide that he really does know, and perhaps to attribute his desire for verification or his reluctance to act on his knowledge to his failure to know that he knows. What would support contextualism is not the (simple) linguistic variability DeRose cites, but a reflective linguistic variability by informed speakers.

What I discern instead is an inclination to reconcile the initially discordant knowledge attributions of different contexts. If, in the more demanding context,

---

[24] I hope this verdict is intuitive. I will defend it and give further examples in Chapter 8.

the speaker says of his earlier attribution in the less demanding context that it was *then* in error, he is reconciling his usage in a way that denies the contextualist his data. Of course, the contextualist can reply that the reconciliation occurs in a fixed, new context. But now the grounds for individuating contexts have shifted. Without a change in what error risks, it is less clear that the inclination to use "knows" is contextual. Moreover, adverting to a new context is a contextualist answer to an objection, not an argument for contextualism.

Despite the contextualist's emphasis on the supposed variability of usage of "knows", I find linguistic behavior pervasively to favor invariance with respect to standards for knowing. Consider this exchange: S_1: "Do you know whether P?" S_2: "Yes I do; P." S_1: "Hold on! P entails Q, which is questionable." Hasn't S_1 disagreed with S_2? Certainly S_1 takes himself to in disagreement. S_1 *claims*, at least by conversational implicature, to disagree with S_2 as to whether S_2's condition is one of knowing. But according to contextualism, the proffered entailment may well (will? What does it take?) change the context. If it does, there is no disagreement as to whether S_2 knows. So the linguistic data do not support a contextual shift.

Consider the reply: S_2: "Since I know that P, Q, although questionable, must be true." Now, for the contextualist, there can be disagreement, but only in the new context, presuming it to be a context that S_1 and S_2 share (and presuming that sharing the context suffices for sharing standards for the application of "knows"). The reply does not *reaffirm* S_2's original conviction in the face of opposition. For it does not use "know" in the original sense; alerted to Q, S_2 has changed contexts. This is going to come as news to S_2, who thinks he is *answering* S_1. What if the reply, instead, is S_2: "Really? Well then, perhaps I do not know that P after all." Now S_2 is *acceding* to S_1's point; S_2 abandons his claim to know. But this is not the verdict of contextualism. S_2 has entered a new context, and his use of "know" in the new context does not contravene its original use. According to contextualism, S_2 has not expressed any change in his position, for it would *still* be correct, in the original context, to claim to know.

I do not see how the linguistic data can be made consonant with this reading. Are we to disagree with S_2, and contend that in supposing his original claim to know to be in error he mistakes a contextual change for a reversal of position? If this is the outcome, I do not foresee S_2's conversion to contextualism. And if contextualism is to be judged as a thesis about how speakers use epistemic terms, this recalcitrance would seem to be a setback.

This criticism is muted, as it is designed for neutrality as to the determinants of context; I intend it to apply to any contextualization of the applicability of "knows". I believe the criticism sharpens if context is identified with the speaker's implicit standards for the correctness of knowledge attributions. For then I do not see how the exchange just imagined can make any sense, whereas it is commonplace to find out what and whether people know by asking them. If the applicability of "knows" is contextualized to the attributer's standards for knowing, then the very intelligibility of such inquiry would seem to require that standards be shared. Otherwise, it is hard to think what S_2 can mean by asking S_1 whether S_1 knows. Surely S_2 is not asking S_1 to judge whether S_1 meets S_2's standards. But S_1's self-ascription of knowledge would seem nonresponsive to S_2's question. Without shared standards, perfectly straight-forward, ordinary conversation becomes an indecipherable mess. With shared standards there is no work for the contextualist thesis to do.

My third of the thoughts I promised about the contextualist's linguistic move is that it raises a problem of self-application. If contextualism, linguistically understood, is *true*, then aren't the conditions for the correctness of what the contextualist says about "knowledge" or "justification" distinctive of the philosophical context in which he labors (perhaps even distinctive of *him*)? How can we assume that what he says correctly describes the behavior of these terms in other contexts (or their use by other speakers)? Why should we think that what he says in his context describes how the use of these terms compares across other contexts? I contend that the use of these terms does not exhibit the variation he imputes, and I do not see how, locked, by his lights, in a context of his own, he can consistently take issue with me. He cannot without paradox claim to know or to be justified in believing that I am wrong.

I myself have difficulty reading "*P* is justified" or "*P* is justifiedly believed" on the model of "I am sad"; it is difficult for me to discern indices in them (other than a pronoun for the believer whom contextual change in this case leaves invariant). And my difficulty only increases with the admonition that contextualism is a linguistic thesis. Therefore, I will revert, for future reference, to the supposed contextualism of epistemic properties. Against all versions of contextualism, however, I contend that the contextualist response to skepticism mistakes a pragmatic for an epistemic difference. A raise in the stakes for being right affects not one's belief, nor one's entitlement to it, nor what it is to be entitled, but rather one's behavior. *Epistemic* justification has no pragmatic dimension.

As I argued in Chapter 5, the contextuality of epistemic practice does not make justification itself contextual. Of course, the evaluation of any belief presupposes other beliefs whose justification is not in question. Of course, one's entitlement to such background beliefs is contextual. Not everything can be in question at once, and what we presuppose and what we question are contextual. But none of this implies that standards of justification are contextual. What is contextual is how high a standard of justification one's justified belief must meet to authorize action.[25] It is standards for action that are contextual, not standards for credence. Beliefs

---

[25] Fantl and McGrath (2002) propose a pragmatic necessary condition for epistemic justification: roughly, a belief is justified only if one's rational preferences are correctly ordered conditionally on its truth. So, for example, I am not justified in believing *P* if, because of what is at stake, it would not be rational for me to act straight away (without further inquiry) on the belief that *P*. This principle simply assumes that it is justification as such, rather than its degree, that rationalizes preferences. Fantl and McGrath make this assumption because the justification they have in mind is justification sufficient for knowledge; if one is justified but does not know, then one does not lack knowledge for lack of justification. They take it for granted that if one knows that *P*, then one's rational preferences are those conditional on the truth of *P*. But suppose that although one knows, one does not know that one knows. One's justification may be sufficient for knowledge without being sufficient to fix one's rational preferences, because it does not settle for one the question as

do not have to be acted on. They can arise simply from the pursuit of incidental interests.

Perhaps, then, the relevance of alternative possibilities that falsify one's belief is contextual simply in the sense of the direction of intellectual attention. A context is just a topic of conversation (or thought). The alternatives to my having hands that I must be able to rule out, as a condition of believing justifiedly that I have hands, are those that happen to come under scrutiny. One does not normally consider skeptical scenarios; they come up only in the (abstruse) context of foundational reflections on the nature of knowledge and epistemic justification. In contexts normal for questioning the truth of paradigmatically justified beliefs, the relevant alternatives are readily checkable. I can find out whether Jones's identical twin is visiting from abroad, whether a mirror is deflecting the candle's light. I can confirm my beliefs as to the locations of Jones and the candle against normal challenges. The justification of paradigmatically justified beliefs is unproblematic unless and until skeptical doubts are raised, which changes the context.

Of course, this means that (lots of ordinary) beliefs are not justified as such at all; they are only justified in a context. In other contexts, they are unjustified. The skeptic may be happy with this result. All he has to do to win is show up. In fact, he needn't even do that, for he is always, implicitly, present. Suppose the context is such that to be justified in believing P I must be able to rule out detectable alternatives Q_i but not skeptical alternatives R_i. Contextualism then denies that I am justified in believing P *tout court*. I am justified only in believing P *as against* Q_i, for I am *not* justified in believing P as against R_i. What this amounts to, so far as I can see, is that the actual object of justification is not P after all, but the weaker disjunction $P \vee R_i$. Skepticism is ubiquitous in the form of an unnoticed disjunct that qualifies all epistemic entitlements.

7.8 Deserved Consideration

But even if contextual justification were thereby to make some kind of headway against the skeptic, I do not think we could avail ourselves of it. The problem is that context is not the relevant standard of relevance. What matter are not what alternatives context makes relevant, but what alternatives *deserve* to be relevant. What *happens* to come under consideration cannot be the proper criterion of relevance, for the simple reason that being justified in one's beliefs is (partly) a matter of *how well* one identifies and scrutinizes alternatives that threaten their truth. I am justified in my beliefs to the extent that my evidence or reasons (if any) are sensitive to ways in which I could go wrong. Evidence that I could have whether or not my belief is true is *weak*. One thing that makes a reason to believe a method reliable *good*, remember, is that the reliability of the method is needed to explain why the reason

---

to whether one does know. It may be irrational to act on P straight away, without further evidence, even if one knows that P, if one does not know that one does.

holds. A reason is good to the extent that it rules out rival explanations. If this is right, then the rivals, or alternatives that I need to be able to rule out, cannot be determined by what I happen to consider.

If the criterion of relevance were just what happens to fall within my purview, then my justification would *increase* as the scope of my attention contracts.[26] The correct connection between justification and inquiry is precisely opposite. What I need to consider and to be able to rule out in order that my belief be justified cannot be *identified* with what I *happen* to consider. For unless it is possible for me to *fail* to consider what being justified *requires* me to consider, justification is achieved trivially; it can have no meaningful measure. What contextualism salvages is not, then, justification at all.

But if a contextualism of this sort looks too weak to deliver justification, it is also, in another respect, too strong. And this latter liability may apply to contextualism very broadly. Once any general criterion of relevance for what must be rejected has been fixed in a context, there is the danger that beliefs that deserve to come out justified will have unrejected relevant alternatives in this context. I do not have to verify that a clock is working to form a justified belief as to the time by consulting it, not in a context in which the clock appears undamaged and its reading is consistent with what I independently know. I have not rejected the possibility that the clock is stuck on a time close enough to the actual time that there would be no evident, independent indication of error, if any. Why is this not a relevant alternative?

In "Elusive Knowledge", David Lewis (1996) says that the reliability of processes that transmit information may be taken for granted. Although he is thinking of perception and memory, not artifacts, we can interpret him as providing a basis for ruling the clock's possible unreliability irrelevant. However, Lewis's criterion is defeated by the very thought that clocks sometimes malfunction. No alternative is properly ignorable, according to him, unless it is actually ignored (by which he means, mistakenly, that the alternative does not fall under one's attention; in fact it *must* come to one's attention to be ignored, for one cannot ignore that of which one is ignorant). It seems evident to me that my belief as to the time is justified without checking the clock's reliability, even if I am occurrently mindful that clocks are imperfect.

Can we keep the presuppositional status of reliability and drop the ignorance condition? Then alternatives actively under consideration are irrelevant, and alternatives unconsidered are relevant. This abandons the contextualism that Lewis thinks necessary to answer the skeptic. He will need a new contextualism that modifies his rule of attention, if his theory is not to reduce to a (vulnerably rough form of) reliability theory.

---

[26] At least, justification would become correctly attributable to me as alternatives escape my attention, if the difference matters. As I have argued against the contextuality not only of the property but also of the concept of justification, and of the proposition that one is justified, I do not think the difference matters.

Peter Klein (2000) proposes an evidential standard for relevance.[27] He thinks that the alternatives that must be ruled out are those that there is some evidence to support. Unless there is evidence for them, we don't need evidence against them. In Klein's example, to justify believing that an animal in the zoo is a zebra does not require evidence that the animal is not instead a painted mule. Might the possibility that the clock is stuck be irrelevant because there is no evidence to support it?

It is intuitive that what requires scrutiny, what is deserving of consideration, is not just any possibility but one that we have some reason to take seriously. But I do not see that Klein's evidential requirement captures this intuition. If someone hands me a utensil which I identify as a spoon, must I not be able to rule out its being instead a fork? Does the justification of my belief that it is a spoon not depend on the adequacy of my reasons—visual, tactile—to eliminate the fork alternative? But there was no evidence that it was a fork. Of course, it is a utensil and forks are utensils. But if this is enough for evidence, then there was evidence that the animal was a mule, for it was an animal and mules are animals. In general, it does not seem that a reason to take a possibility seriously need take the form of evidence for its truth. Why can't the importance of its consequences suffice? In this case the consequences could be dire. I might have just been served a crème brûleé.

Moreover, depending on how loosely we construe evidence, it is unclear that Klein's requirement is not satisfied in the clock case. Clocks do get stuck; indeed, it is common for public clocks to malfunction. Stuck clocks are correct twice a day; they are close to correct more than that. A world in which I am wrong to trust the clock does not seem that far off, the practical consequences of misjudgment can be made as dire as one likes, and we can stipulate that my getting the time right is a matter of immediate interest. Yet I have a justified belief as to the time without verifying the clock's reading or dismantling it to assess the state of its mechanism. Sometimes justified belief does not require investigating even relevant alternatives. I do not know of a standard of relevance that enables contextualism to get justification right.

7.9 Plausibility

None of the anti-skeptical measures I have considered protects the justification of ordinary beliefs or ordinary knowledge against the possibility of skeptical scenarios. Each measure must presuppose that skeptical scenarios do not apply to satisfy the

---

[27] Klein's purpose is not to defend contextualism, but to answer the skeptic on independent grounds: since there is no evidence to support skeptical hypotheses, the justification of ordinary beliefs does not depend on evidence against such hypotheses. Thus, independently justifiable, ordinary beliefs are then eligible to be turned against skepticism. This later move is congenial to my own defense of justificatory inference. However, Klein stops far short of my general endorsement of justificatory inference, for he rejects CJC on which I have argued that justificatory inference depends. (See his 1981, p. 79.) Klein is a good example of an epistemologist who wants reasoning to be justificatory but cannot fully make it so, because he cannot bring himself to face and accept the consequences of CJC.

conditions it requires for ordinary justification or knowledge. Regardless of whether or not inference is justificatory, as my theory requires, ordinary beliefs are justified only if the rejection of skeptical scenarios is justified. Whether skeptical scenarios are rejected by inference from ordinary justification, as on my theory, or as a precondition for the justification that another theory constructs for ordinary belief, they must be rejectable. The generic problem in attempts to show otherwise is that the accessibility of subjunctive possibilities varies across worlds that are not discriminable, with skepticism in the running. The standards proposed to discriminate among worlds as to relevance beg the question against skepticism by presupposing that ours is not a skeptical world. As skepticism denies that the presupposition is justified, an answer to skepticism cannot depend on making it.

I surmise, however, that when standards like proximity or similarity or realism or intellectual interest are proposed, what is really going on is something different. The intuitive appeal of these stratagems, it seems to me, is *entirely* a matter of simple plausibility. It is plausible to suppose one to lack hands as the result of an accident. It is readily understandable how that could happen. The operative basis for restricting the range of ways of going wrong that justification requires one to check, whatever it is called, amounts to plausibility. Certainly plausibility gives the *same* results in the ranking of relevance as any of the more complicated conceptual machinery philosophers have deployed. And unlike this machinery, plausibility *works*; it does not beg the question against the skeptic. Descartes was forthright about the implausibility to which the search for a systematic basis for doubt drove him. It is only charitable, it seems to me, to *interpret* contextualism and possible-worlds metaphysics as (roundabout) appeals to plausibility.

But if plausibility is an acceptable criterion of relevance, skepticism loses. Skeptical scenarios are as far-fetched as it gets. Who is this mad scientist, anyway? Why does he do it? How does he do it? What is his interest in me? How can he know so much about my earlier mental life? Such explanatory lacuna are the essence of implausibility. They induce the dissatisfaction we experience at brute stipulation. To fill them in coherently requires suspension of a natural skepticism about skepticism.[28]

So we come to this situation: Either plausibility is an acceptable standard and we *are* justified in rejecting skeptical scenarios; or plausibility is unacceptable, in which case skeptical worries arise immediately at the level of ordinary beliefs.[29] I am not even justified in believing I have hands (let alone that I am not systematically deluded) if the *immensely* greater plausibility of the belief that I have hands over any skeptical scenario, and any philosophical principle that can underwrite a skeptical

---

[28] I frequently hear the Matrix movies cited to make sense of extreme skeptical scenarios. They are even referenced in syllabi for introductory philosophy courses, presumably to connect with student interest. I can only say that these movies, though intermittently entertaining, do not make coherent sense to me, and I have yet to find anyone able to explain them.

[29] Perhaps some skeptical scenarios are not extreme, but are more plausible than the classic ones that have traditionally exercised epistemologists. Maybe the matrix does make sense, and I just don't get it. As it is not my mission to refute skepticism, I will not argue the point.

scenario (like DP), does not count. In *neither* case is there any call to fault, nor any advantage in faulting, my transmission principles for justification. No theory of knowledge or justification, with transmission or without, shows how ordinary propositions about one's immediate environment can be known or justified if propositions denying that one is victimized in a skeptical scenario cannot be known or justified. And so I conclude that nothing is gained against the skeptic by opting for a theory of justification or knowledge that suspends or restricts the transmission of these epistemic properties through truth-preserving inference.

For my part, I am happy to dismiss skeptical worries on the grounds of plausibility alone. There is always a choice: ordinary beliefs are justified and skeptical arguments to the contrary are (somehow; there are lots of arguments) faulty, or the arguments are to be trusted over ordinary beliefs. Given this choice, if plausibility counts for anything then skepticism loses.

I am not alone in my satisfaction with this argument from plausibility. This way with skepticism is reminiscent of G.E. Moore (1959). Moore formulates the choice as between arguments: Either I don't know I'm not victimized and infer that skepticism is true (I don't know I have hands); or I do know I have hands (skepticism is false) and infer that I am not victimized. The appeal to relative plausibility in choosing against the skeptic is Moorean. It is common to dismiss Moore's way with skepticism for failure to take skeptical arguments seriously. Those who conceive the mission of epistemology to be an adequate response to skepticism are invited to interpret my reliabilist argument as providing a theoretical basis for a Moorean response. If ordinary beliefs are justified and justification answers to my reliabilist standard, then the rejection of skeptical scenarios must also be justified. This is a serious reason to reject them.

Chapter 8
Tracking and Epistemic Luck

8.1 Attractions of Tracking

According to my theory, paired down, a method of belief-formation is justificatory if it is counterfactually sensitive to falsity. The method is not required to be sensitive to truth. Thus, mine is not a tracking theory; belief-formation does not covary with, or track, truth-value.[1] This is unfortunate with respect to our epistemic goal. But it was to be anticipated. It reflects the greater relative importance of falsity-aversion over truth-acquisition within this goal. Believing falsely is epistemically worse than failing to believe truly. No general method can guarantee the acquisition of truth and nothing but.

There is further misfortune if one counts it an attraction of tracking to disqualify beliefs produced by otherwise reliable methods that are epistemically compromised by unjustified beliefs about what makes these methods reliable. With tracking, beliefs obtained by consulting an encyclopedia come out unjustified if one's intention is to consult an encyclopedia blessed by the Pope. For this method will fail to yield truths under the counterfactual supposition that the only encyclopedia available is one the Pope has not blessed.

I have argued, however, that beliefs thus dependent upon features of a method irrelevant to its reliability should not be disqualified. So long as one would not use unreliable sources the Pope *has* blessed—so long as the Pope's blessing is not sufficient—one's method is reliable. One may justifiedly believe a method reliable despite having false and unjustified beliefs about what makes it reliable. One's justification will consist in *other* beliefs one has about the method, and will not be defeated by one's additional, adventitious confusions. For otherwise, precious little justification may survive. It is too much to ask that our theories about what makes

---

[1] There is some terminological leeway here. I do have what can be considered a tracking condition. I deny that mine is a tracking theory to emphasize my rejection, anticipated in Chapter 1, of a further tracking condition. What I mean by a "tracking theory" of justification or knowledge is one that requires belief-formation to be sensitive to truth as well as to falsity. As explained in Chapter 1, this is how Nozick originally used the term "tracking". He introduced it (1981, p. 178) to signify *joint* satisfaction of the two conditions. My terminology respects the fact that according to the inventor of tracking theory, mine is not such a theory.

J. Leplin, *A Theory of Epistemic Justification*, Philosophical Studies Series 112, DOI 10.1007/978-1-4020-9567-2_8, © Springer Science+Business Media B.V. 2009

our methods reliable be inerrant; it is enough that they are substantially true. Despite the causal inefficaciousness of its gratuitously expensive pedigree, Bayer aspirin does relieve headaches.

Tracking's original (originating) attraction was to disqualify beliefs induced by external manipulations that could not be performed unless the beliefs they induced were true. Nozick imposes the condition of subjunctive sensitivity to truth, and, on the basis of this addition to his theory, introduces the term "tracking", specifically to disqualify such beliefs from constituting knowledge (1981, pp. 175–178).[2] The mad scientist cannot induce in me the belief that he is inducing a belief in me unless this belief is true. Nevertheless, this belief does not seem to be justified, because he could have made me believe that the belief he induces has some different etiology. Indeed, it is normally part of a skeptical scenario that induced beliefs do *not* reveal the victim's true position, for they are supposed to be false despite the victim's (apparent) evidence. Tracking solves the problem, because the victim's beliefs arise in a way that owes nothing to their truth (if any).

Of course, this is not my solution. My theory solves the problem by distinguishing beliefs induced externally from beliefs the believer himself forms by intentional application of a method. The former are unjustified even if the mechanism of induction is nomically related to the semantic content of the beliefs induced, for the believer is not forming them. Nor does the believer believe them justifiedly, although he does so blamelessly, if no apparent grounding accompanies their imposition.

8.2 Against Tracking

Attractive or not, tracking is unacceptable as a condition for justification.[3] One reason is a point made earlier. A method of belief-formation may be used with insufficient skill or care to reach a result obtainable by it. In general, that a belief is true is insufficient to guarantee that a reliable method will deliver it, even under normal conditions.

---

[2] Nozick says that a person in a tank who is manipulated into believing that he is in the tank satisfies subjunctive sensitivity to falsity, but not truth. Since the person does not know, sensitivity to truth must be added to the theory. (As noted below, no such addition is called for on my theory, since my reliability condition is not satisfied.) Of course, Nozick's developed theory has more general motivations. Tracking is supposed to generalize the causal connection between a truth and a belief that is often felt to make the difference as to whether the belief is knowledge. And Nozick says (1981, pp. 170–171) that his motivation for investigating this connection was to show how freedom of action could be protected against the causal determination of action.

[3] Does tracking fare better as a condition for knowledge? Does the fact that Nozick is concerned with knowledge rather than justification account for his imposition of the stronger condition? Recall from Chapter 7 that Pritchard follows Nozick in adopting a tracking condition for knowledge. Tracking does fail in some cases where justification succeeds but knowledge fails. I consider such cases below in connection with the epistemic role of luck. But there are also cases of knowing that do not satisfy tracking, and I shall argue below that knowledge does not, in general, require tracking.

This point is difficult to formulate because of a tension in the notion of an "undelivered belief"; should we not speak instead of a true but unbelieved proposition? An easy way to make the point is to suppose that one agent forms the belief while another, for want of skill or tenacity, does not, although his method is the same. The grandmaster finds mate in four, but the master does not see it; there need be no difference in what they intend to do as they seek the solution. We may extrapolate, however, to the case in which no one reaches a truth that is nevertheless potentially reachable by a method that inquirers employ. In stating my theory, I was able to let the supposition that a belief has been formed identify the method(s) of belief-formation on which the belief's justificatory status depends. It is the method intentionally used in forming the belief. With the extrapolation, it becomes unclear what method or methods are at issue.

Intuitively, a method is applicable to propositions of a certain kind, subsumable under a general description of a domain of subject matter. This is not to endorse the "epistemological realism" that Michael Williams (1991) cogently attacks. We do not rely on the semantic content of beliefs to identify the domains to which they belong. The domains I am talking about are ranges of applicability for intentionally usable methods. But the relevant notion of applicability is ambiguous. The domain with respect to which a method is reliable could (unfortunately) differ from that to which it is applied. Some methods should not (prudentially) be used at all, because they are reliable only with respect to domains much better investigated by other methods, or better left uninvestigated. Reading tea leaves is a reliable method of ascertaining the distribution of tea leaves about the bottom of one's cup, but this domain is devoid of interest. Crystal ball gazing is reliable with respect to the presence of light-emitting objects about the room, but these are better identified by looking at them directly. Wishful thinking is reliable with respect to self-fulfilling propositions that believing makes true, but this is not the truth-condition for these propositions that interests us. In identifying the domain to which a method is "applicable", we assume the concordance of reliability with use.

Reading a thermometer establishes judgments of temperature; mathematical deduction from axioms and theorems establishes judgments of theoremhood. Where the particular belief a method is to deliver is unspecified, the method must be identified independently, and its reliability must be relativized to a domain. The broader or vaguer the domain's specification, the less clear a notion reliability becomes, for a method highly sensitive to falsity in some (easy) cases may be less so with respect to others of the same general kind. Sight reliably detects objects of ordinary size in one's vicinity, but the table under the lamp in front of one better than the chair recessed in an alcove.

I further consider the notion of the range of a method's reliability in Chapter 10. My point at present is that a method may be differentially sensitive to falsity and truth within the same domain under the same (normal) conditions; it does not steer one wrong, but only with diligence and, perhaps, luck, if at all, does it steer one right. It delivers *some* beliefs, for it is a method. But propositions in the same domain as these are more resistant. Some truths within a domain with respect to which a method is reliable may remain undiscovered however well the method is applied.

This can happen in two ways: either the truth of these truths is unrevealed despite the application of the method to them, or the method fails even to identify them for consideration.

Perhaps no one can prove Goldbach's conjecture, though it is a theorem. If a mathematician proves Goldbach's conjecture he knows (and justifiedly believes) it is a theorem, even though he could, despite his efforts, have failed to prove it and so have failed to believe it to be a theorem. Perhaps there is a mate in four that no one finds. The grandmaster, by analyzing the position, knows he has mate in four. But it is not the case that were he to have mate in four and analyze the position he would then believe he has mate in four. He could easily have failed to find it and perhaps often does fail in positions of comparable complexity. These are ways for reliable belief-formation to violate tracking. One can achieve knowledge and justification with respect to a belief that one could have failed to form, even though it were true and one applied the very method by which knowledge and justification were in fact achieved. So neither knowledge nor justification requires tracking.

In enabling us to identify the methods of belief-formation used by differentially successful inquirers, the intentionality component of my theory produces counterexamples to tracking. At a magic show I watch a magician perform a card trick. A volunteer from the audience has selected a card at random, revealed it only to the audience, and supposedly returned it to the deck and reshuffled. Everyone in the audience anticipates that the magician will identify the card; all watch intently to figure out how he does it. What distinguishes me is an inadvertent glance away from the area to which the magician has cleverly directed attention, occasioned by a twinge in my left ear. I alone notice the magician deftly purloin the card and secrete it up his sleeve; it has not been returned to the deck after all. I thereby form the true belief that he will scan the remaining deck and identify the selected card by process of elimination. (Magicians are not psychics; they are skilled, after all.) The inadvertence of my glance shows that nothing is distinctive about the method I intentionally use. Scrutinizing the magician's motions produces a true belief in my case but not in others.

Without reference to what I intend, it is unclear that my method cannot be made distinctive by incorporating additional features of the situation into it. The extra glance, the twinge, are components of what happened in forming my belief without analogues in the case of others in the audience. But there is nothing abnormal about them. That they are not components of any method I intentionally applied establishes that my justified belief need not have been produced, even though it were true and I were to apply the very method that in fact justifies it.

Nozick discusses cases similar to my magician case. Most similar, but relatively neglected, is his case of the escaping criminal, in which an inadvertent glace is the basis of belief. The Jesse James case is more discussed. In it, what is noticed, rather than the action that brings this to notice, is happenstance (1981, p. 193). Without the intentional component of my analysis of reliability, it is difficult to decide what to make of such cases. How much of what happens in them is constitutive of one's method? How much must be held fixed across counterfactual situations in which the method is used but no belief or a false belief is formed by it? To defend the necessity of tracking for knowledge against the Jesse James case, Nozick must include the particular visual imagery obtained by inadvertence or happenstance as part of the method. But then we seem to get tracking in cases where there is no knowledge. To defend the sufficiency of tracking for knowledge against such cases, Nozick identifies methods much more flexibly. So, for example, one does not know Judy from Trudy, if one's correct identification of Judy depends on one's accidentally coming to believe Judy to have a distinguishing feature that she accidentally comes to possess. Here the feature does not count as a constituent of one's method.[4]

This expediency opens Nozick to accusations of inconsistency and to the criticism that his theory fails to provide a clear criterion for the identification and individuation of methods.[5] The critics want to refute Nozick's theory; it is not to their purpose to decide what is the right way to achieve a consistent theory. My theory does not leave this matter in limbo. We get justification in the Jesse James case, as in the magician case, because a justificatory method is not required to deliver truth. The identification of Judy is unjustified, because there the method does not satisfy reliability. Experiential features of a case of belief-formation that are obtained by inadvertence or happenstance are definitively not constituents of the method one intentionally uses. Therefore, justification on my theory does not require tracking. It does not require subjunctive sensitivity to truth, even under normal conditions.

8.3 Fortunate Ignorance

Getting the magic trick is an example of epistemic luck.[6] My justified true belief is fortuitous fortune. I do not expect the legitimacy of luck in this case, its innocuousness with respect to epistemic status, to be disputed. For what is lucky is one's special knowledge, and knowledge is, presumptively, good. One knows that the mystery card has been separated from the deck, and this intelligence enables one to discern the trick. The legitimacy of luck *is* disputed, however, where what is lucky is one's ignorance.[7]

---

[4] This example originates with Goldman (1976).

[5] Robert Shope (1984) and Graham Forbes (1984) develop these criticisms in detail.

[6] Duncan Pritchard (2005) identifies numerous complexities in the usage of "luck". In particular, it sometimes means just fortune, and sometimes means fortunate fortune. I will generally be using it in the latter sense, but my purposes do not require an analysis to fix a consistent usage.

[7] The philosophers I have in mind as disputants—e.g., Gilbert Harman (1973), Simon Blackburn (1984), David Lewis (1996)—are talking about knowledge, not justification, so it is strictly open to them to agree with me. But, with the exception of Blackburn, these philosophers leave it simply intuitive that knowledge disallows fortuitously fortunate ignorance. (Blackburn attributes the intuition to a supposed social function of knowledge that strikes me as gratuitously derivative, of which more later.) If sound, this intuition would seem applicable to justification as readily as to knowledge, and it must therefore be rejected. Those sensitive to the symmetry I am about to recommend will not share it.

I can only suppose that this is because ignorance is, presumptively, bad. I do not find this to be a very good reason. It seems to me that there is an important symmetry in the epistemic role of luck, and that if justification is to be granted when dependent on lucky knowledge, it must also be granted when dependent on lucky ignorance. For what if, instead of fortuitously noticing something others miss and discerning the trick in this way, I fortuitously miss what others are manipulated into noticing and am thereby undeceived? The magician directs attention to a prop that misleads the audience, while I, wincing at the twinge in my ear, close my eyes momentarily and am not taken in. Surely my resultant conclusion as to how the trick is done is as justified in this case as in the former.

One might think of cases in which the fact of one's ignorance outweighs in importance the fact that what one is ignorant of is best unknown, or, if known, best ignored. One's ignorance compromises one's authority or trustworthiness in a general way, even though it is fortunate on this occasion. Ignorant of van der Waal's equation for gases, I apply the simpler, idealized Boyle-Charles law and get the right answer. Had I known better, I could not have gotten the answer, for the additional quantities needed to apply the deeper theory were not determinable. Despite its present benefit, my ignorance is, on balance, disadvantageous; I am less reliable than is one more knowledgeable.

But why are such cases more telling than cases in which we regard ignorance as an advantage? Often we seek a fresh mind, unconstrained by ingrained ways of thinking and able to tackle a problem free of unrecognized and possibly misleading preconceptions. Experience, claimed Nixon, qualified him over Kennedy. But experience of the wrong kind, in that case corrupt and exploitive, is a *dis*qualification. Whether, for the purpose at hand, knowledge is better than ignorance depends on what kind it is. In forming beliefs, ignorance can be epistemically beneficial and knowledge deleterious, as readily as the reverse.

One obtains a class of counterexamples to tracking by insulating the believer from misleading but justifiedly believable defeaters of his belief. We get justification (and can get knowledge as well) in these examples, although only luck protects the believer from conditions in which he would not have held the belief despite its truth. These conditions can be made proximous by anyone's standard.

CNN reports the capture of a wanted terrorist: alerted by a CIA informant, marines assaulted a resort where the terrorist was meeting with his financiers, and took him into custody. Then an electrical storm knocks out power to my house and I miss CNN's subsequent retraction, in which the captive is identified as a double the terrorist has often used to evade his pursuers. The retraction is in error, however. The double was to have replaced the terrorist on this occasion, and CNN issued the retraction when this plan was leaked. But at the last minute, the double fell ill and the terrorist had to appear after all. My belief that the terrorist is in custody is justified (and true), but my method, watching CNN, would not have sustained it under the counterfactual supposition that my power stays on. It is just lucky (epistemically, at any rate) that the power went out. And of course, it is unnecessary to the example that the defeater be false. It could just be the information that a double has often been taken for the terrorist, which is true and sufficient to undermine, incorrectly, one's confidence in CNN's initial report.[8]

If one responds to the example by denying justification, one must then either deny symmetry or deny that serendipitous knowledge can justify, as in the original magician case. Neither consequence makes sense to me. If their dubiousness is insufficient deterrence for you, I offer the following remedy. There is indeed an intuition that justification should not be a matter of luck. Believing justifiedly *contrasts* with believing truly by luck, although it could be just luck that one's justified belief is true. This intuition, however, is coarse. It does not identify where, exactly, the involvement of luck becomes objectionable.

8.4 Roles for Luck

I suggest that there are different roles for luck to play, and not all are epistemically deleterious. Some are altogether innocuous, and some affect the justification of belief and believing differently. The reliability of a justificatory method should not be happenstance. The worry that reliability can obtain by accident drives some philosophers to internalism, and me to distinguish a belief's justification from the believer's justification. If I am nomologically insulated from deceptive evidence and ignorant both of the evidence and of my invulnerability to it, then my method may be reliable by luck. Suppose other networks that I would trust as readily as CNN misreport the news, but broadcast on frequencies that my obsolete television cannot receive. CNN is the only news channel I can access. Then listening to television news (that it be CNN, in particular, is not part of what I intend) is for me, by luck, a reliable method of forming beliefs. But lacking good reason to believe it reliable, I am not justified in holding the beliefs thereby justified.[9]

By contrast, a method that is not reliable may, by luck, be applied only when accurate. A clock works properly on alternate days. (To save electricity, it is unplugged and plugged back in on a 24-hour cycle, but I do not know this.) My method of forming beliefs about the time is to consult this clock, but I (happen to) use this method only on days the clock is running. Though inerrant, my method

---

[8] This example resembles Gilbert Harman's presidential assassination case (1973, pp. 143–145). What suggested the example to me was a CNN report, prior to the 2003 Iraq war, that Saddam Hussein routinely used doubles to replace him at public events. In any case, the use Harman makes of his example is just the opposite of the use I shall make of mine. I contend in Harman's case that the believer is not only justified but knows. If the information that undermines one's true belief were true, despite the truth of the belief it undermines, then I would agree with Harman that one does not know. I shall presently be discussing examples of this kind in connection with the (supposed) "safety" condition for knowledge.

[9] My theory delivers the same verdict in Goldman's example of a stranger who smiles just in case he wins the lottery (1986, p. 46). One believes the stranger to have won solely on the basis of seeing him smile. This is a poor example for my purposes (and Goldman's), because there is no plausible account of why the believer would invest such significance in the stranger's smile (let alone why winning the jackpot rather than the lottery would fail to elicit a smile).

does not satisfy the counterfactual standard for reliability. My beliefs as to the time are unjustified, although, lacking reason to suspect the clock's vicissitudes, I may be justified in holding them. Similarly, I am justified in believing the readings of a thermometer which I assume correctly to be in good working order, but which I selected at random at the store from a box of new thermometers all the rest of which are broken. My intended method—not to use this particular thermometer but just to use a new one from the store—is by luck inerrant though clearly unreliable. So the resulting beliefs are unjustified.[10]

Then again, it might be a matter of luck whether one believes what a method, reliable not by luck, delivers. For it might be happenstance that one is exposed to or protected from abnormal conditions in which one's reliable method misinforms. The luck of the power-outage shields me from the misinformation that it was not in fact the terrorist who was captured, but his double. The temperature was below 50°C, the limit of my thermometer's range of accuracy. Somewhat later the temperature climbed above 50°C, but my thermometer climbed only to 50°C. By luck I didn't consult it *then*. The beliefs I obtain consulting the thermometer within its range are justified and I am justified in holding them, although that the temperature be within this range is no part of what I intend. I know nothing of the instrument's limits; I just look at it. Nor could I, were I to know the limits, confirm that conditions are within them. For I have no more sensitive thermometer, with respect to which, in any case, the same problem would recur at higher temperatures.

But for luck I could, by my reliable but uncritical trust in the thermometer, come to false beliefs. This role of luck, it seems to me, is unavoidable but innocuous. So long as one grants the consistency of justifiedness with falsity, one must grant that it could just be luck that some false propositions are not justifiedly believed along with the true ones that are justifiedly believed. For one grants that there are conditions under which a justificatory method will err, and there can be no general protection against these conditions.

One can take precautions against abnormal conditions, and improve one's degree of justification in so doing. But there are no guarantees. Justification can (does, on my theory) require that there not be reason to believe that such conditions prevail, but it cannot presume that where they do prevail there need be reason to believe they do. Thus, I deny that an instability or fragility in normal conditions, their

---

[10] Goldman (1986, p. 45) has an example in which a parent takes a child's temperature with, by chance, the one good thermometer in a medicine cabinet loaded with thermometers stuck on the child's actual temperature. As the parent's intended method is evidently to use just any of the thermometers, it is unreliable despite the fact that the belief it generates would have been true regardless of which thermometer he selected. The method is inerrant only by luck, for it is only luck that the method does not generate false beliefs; it would do so were the child's temperature different and a different thermometer selected. Therefore the parent's belief is unjustified. However, I am also unprepared in Goldman's example to grant justification to the parent, for I am unsure what he is doing with so many thermometers in his medicine cabinet and how this affects his belief-formation processes or reasons for trusting them. For example, if his preferred method is to use several thermometers and believe the results only if they agree, then by hypothesis he is on this occasion resorting (is he pressed for time?) to a method that he does not trust.

vulnerability to an undetectable change into abnormal conditions in which a method errs (supposing, for the sake of argument, that such an arrangement would not itself constitute abnormality), subverts justification.

By distinguishing normal from abnormal conditions, the justification of beliefs from the justification of the believer, and intentional belief-formation from belief-induction, my theory accommodates (much of) the intuition that luck is incongruent with positive epistemic status.

8.5 Safety Versus Sensitivity

In fact, respect for this intuition (is one thing that[11]) prevents me from venturing an analysis of knowing, as follows: a justified true belief that a person is justi-fied in believing is known; if P is true, reliably produced, and believed for good reason to be reliably produced, then the believer knows P. The difficulty is that these stipulations do not prevent P's truth from being luck. For, conditions could be abnormal, and P true anyway. Then P would be believed even if false, but not under normal conditions. To get knowledge we must stipulate that conditions are normal, which makes the truth condition for knowledge redundant. And then the problem is that one's method might be reliable only by luck despite one's good reasons for believing it reliable. Does one know the time if one consults a clock that the mad scientist, at whim, has just restored to accuracy? More generally, what makes one's method reliable could be unrelated to one's reasons for believing one's method reliable, despite the goodness of these reasons. Then one justifiedly believes a true and justified belief, but does not know.

Should we then require that one's reasons and the reliability of one's method be related? How, exactly? It does not look promising to relate them by requiring that one believe the method reliable for the *right reasons*. This is an elusive notion. There may be lots of reasons that vary in their pertinence and degree of accuracy. One cannot, in the best case, expect one's reasons to guarantee that one's method is reliable. And we have seen that one can have good reasons to believe a method reliable while seriously misunderstanding what makes it reliable. One can know in virtue of the reliability of one's method and one's reasons to believe the method reliable, while misjudging the sources of its reliability. One can know by looking despite holding a false theory of vision. Galileo learned a lot about the heavens by telescopic observation, despite his ignorance of geometric optics.

Should we instead try to eliminate luck by concluding that knowing P requires not believing P if P is false, under *any* conditions? I do not think so, but this is a conclusion that many philosophers have almost reached. There are but two steps

---

[11] I am unconvinced that knowledge requires belief. As one can act in ways that one knows better than to act, so, I suspect, one can believe what one knows better than to believe. (Chapter 9 will provide an example.) One may believe out of hope, fear, or indoctrination, despite one's cognizance that these influences are not truth-conducive. One believes, knowing that one is wrong. But one does not believe believing that one is wrong.

separating it from the requirement of *safety* that many philosophers advocate for knowledge (Williamson, 2000; Sosa, 1999, 2000). Step one is to contrapose: knowing P requires that P is true if believed, under all conditions. This protects one's belief against bad luck; it provides an epistemic kind of insurance. If conditions were different but one still believed P, P would still be true. Step two is to qualify the universal quantification over conditions. Knowing cannot literally require that under *no* conditions would one believe falsely, for then there is no knowledge. Rather, safety's advocates think that quantification implicates a restricted domain; conditions outside this domain are not included among *all* conditions. The conditions that count are accessible in virtue of bearing a proximity relation to actual conditions.

With these two steps we get the safety condition for knowledge: if one knows that P, then any sufficiently proximous counterfactual situation that preserves one's belief that P also preserves P's truth. The advocates of safety think that these are big steps. I disagree. I think that what they deliver is essentially the requirement that one would not believe P if P were false. This requirement is too strong both for knowledge and for justification.

One reason to imagine that safety lies at an epistemically significant distance from counterfactual sensitivity to falsity is the conventional wisdom that subjunctive conditionals do not contrapose; that is, they are not equivalent to their contrapositives. Safety's advocates propound it as an *improvement* over its contrapositive, which they label *sensitivity*, contending that safety appropriately delivers knowledge or justification that sensitivity inappropriately blocks. In particular, say advocates of safety, ordinary knowledge would be false under skeptical scenarios but would still be true if still believed. Naturally (given the argument of Chapter 7) I am going to dispute this, but what, first, about the underlying claim of nonequivalence[12]

Subjunctive conditionals carry presuppositions as to how things are. The truth-value of a subjunctive conditional depends on what is presupposed. Contraposition can be violated if the supposition that the consequent of an accepted conditional is false alters these presuppositions. But does it? I find many supposed counterexamples to contraposition unconvincing, because in them it takes more than the supposition that the consequent is false to alter presuppositions. Some ancillary shift of conversational context is doing the work. Had I captured your queen I would have lost (for the queen's exposure was a trap). One is reluctant to say that had I won I would not have captured your queen. For capturing the queen is a way to win (forgetting that this was a trap). In effect, it is not the contrapositive itself that is false, but the contrapositive with a presupposition-shifting clause added to its antecedent.

---

[12] In the standard semantics, the counterfactual from P to Q fails to contrapose if some P&Q world is closer to the actual world than is any P&~Q world, but some P&~Q world is closer than a ~P&~Q world. I am not going to assume the standard semantics, because there are unresolved counterexamples and the proximity metric is disputatious. Lewis's responses to these problems (1973) seem to me ad hoc and stipulative. Also, as explained in Chapter 7, I do not want to assume an ontology of possible worlds.

Ernest Sosa (1999, p. 152) uses an example that builds a presupposition-shifting clause into the consequent position: "If water flowed through your kitchen faucet, it would *not* then be the case that water so flowed while your main valve was closed". The example suggests a formula for generating counterexamples to contraposition: repeat the antecedent within a conjunction that the consequent denies, where the remaining conjunct defeats the first. Then the contrapositive is false because the negation of the consequent automatically makes the antecedent true. However, the conditionals this formula generates are at least bizarre, and unidiomatic; they are not obviously true.

Perhaps Sosa thinks his conditional is true because it is true that if water flows the valve is *not* closed, and also true that if the valve is not closed then it is not closed while water flows. If this is Sosa's reasoning, he is evidently assuming transitivity for subjunctive conditionals. However, transitivity fails: If I were to release the hammer, it would fall. If the hammer were to fall, it would not be the case that the gravitational force on the hammer was counterbalanced by an upward magnetic force. But it is not true (intuitively[13]) that if I were to release the hammer, then the gravitational force on the hammer would not be counterbalanced by an upward magnetic force.

Although I find Sosa's construction unconvincing, it suggests to me that there should be more tractable examples to the same effect. Rather than repeating the antecedent, the proposition that the consequent negates should carry an implicit commitment to the antecedent. This obviates any reliance on transitivity. Consider: If I were to build a new house, I would not employ the builder my cousin used. If I painted my house again, I would not use blue paint. If I were ever to return to Vincenzo's, I wouldn't order cannelloni.[14]

---

[13] Maybe this does come out true on a possible-worlds semantics that measures proximity probabilistically. If so, I think that is trouble for such semantics.

[14] It might be thought that contraposition fails unequivocally if pronouns are introduced into the consequent position: If I painted my house again, I wouldn't paint *it* blue. But the contrapositive of this is unclear, because the reference of the pronoun, occurring in the antecedent position, is not fixed. It may be better to treat such constructions as existentials than as conditionals. A similar objection applies to use of definite descriptions; e.g., "If I were to dine again at Vinzenzo's, I wouldn't order *the* cannelloni". Moved by contraposition to the antecedent position, the definite description has no clear reference. Considered independently of the original conditional, I simply do not understand the supposition that I order *the* cannelloni. Instead of using a pronoun or definite description, one might repeat the noun in the consequent position. But then the result is unidiomatic, and, pursuant to what I suggest below, is not properly interpreted as conditional in form. A similar objection applies to David Lewis's (1973, p. 35) example: "If Boris had gone to the party, Olga would still have gone." Says Lewis, if it was just to avoid Olga that Boris did not go, then it is not the case that "If Olga had not gone, Boris would still have not gone". Contraposition is a *syntactic* operation, and Lewis's surreptitious relocation of the key term "still", which makes all the difference, is impermissible exactly as the interchange of nouns and pronouns is impermissible. To contrapose, one negates and interchanges the components; further tinkering is tolerated only to the extent necessary to restore grammaticality. One can claim that if a conditional is clear but its contrapositive is unclear, this is enough to render them nonequivalent. I maintain, instead, that it

It remains possible in these cases to deny that the contrapositives are false. If I employed my cousin's builder it would not be to build a new house, since I would not employ him to do that. If I used blue paint, I would be painting something else, for I would not again paint my house that color. If I were to order cannelloni, it wouldn't be at Vincenzo's. However, it also seems reasonable to deny that these examples contrapose. If I used blue paint this would be because other colors are unavailable, contrary to the antecedent's presupposition that I have a choice. So it is not the case that if I use blue paint, then I am not painting my house. If I employed my cousin's builder, I would have changed my opinion of him. If I order cannelloni, Vincenzo's will have changed chefs. I think that cases fitting my formula can go either way, and this is reason enough to rule against contraposition.

But although subjunctive conditionals do not, in general, contrapose, I claim that those that do not contrapose have idiomatic equivalents that do contrapose, if these idiomatic equivalents are themselves of conditional form. An idiomatic equivalent expresses idiomatically what the speaker means. So, if contraposition fails for the builder example, this is because what I mean by the original conditional is that I will not employ my cousin's builder, whether or not I build a house. I am equivalently saying that I won't employ him at all. The conditional form of the original is merely emphatic. In the painting example, I am saying that I prefer another color. And I do not care for Vincenzo's cannelloni. Idiomatic equivalents of Sosa-type cases do not repeat the antecedent; the repetition is unidiomatic because gratuitous. So they do contrapose: if the valve were closed water would not flow.

Counterexamples to contraposition depending on necessities also succumb to my claim. Is it true that if $2 + 3 \neq 6$ I would not believe that $2 + 3 = 6$, but false that if I believed that $2 + 3 = 6$ then $2 + 3 = 6$? If the first conditional is true, this must be because I do not believe that $2 + 3 = 6$ period; the antecedent does no work. Certainly, there is no general connection between suppositions as to the truth-values of necessities and what one believes about them. But if one does not believe that $2+3 = 6$ under any conditions, then how is the antecedent of the second conditional to be entertained? Clearly it, too, does no work. If these conditionals make sense at all, the sense they make is not of conditional form.

If my claim is correct, then the failure of contraposition for subjunctive conditionals is not a good reason to distinguish safety from sensitivity as a significantly different condition. And in my view, there is no epistemically important difference between them. My reliability condition contraposes: if under normal conditions the method I intentionally use in forming a belief would not lead me to form the belief were the belief false, then under normal conditions were I to form the belief by intentionally using this method, the belief would not be false.

Of course, my reliability condition is not relativized to a range of counterfactual situations. But if it is necessary to specify the range of counterfactual situations in which a belief's falsity prevents one from holding it, then it is equally necessary to

---

is then unclear *whether* they are equivalent, and that the unclarity of the contrapositive should be resolved in favor of equivalence on general principles.

specify the range of situations in which sustaining the belief ensures its truth. The latter project seems to me as problematic as the former. I argued in Chapter 7 that with skepticism at issue, there is no way to narrow these ranges without begging the question. Safety does not enable one to know that one has hands, if the world in which one truly believes this is one to which a skeptical world is proximous. If this belief is insensitive because held in a world in which it is false, then it is also unsafe if this world is among those in which it must be true if believed.[15]

Short of skepticism, the specification of ranges of proximous worlds will depend on indefinitely extendable independent assumptions about the actual world. Such further assumptions may distinguish the ranges, but they will not, in general, render them disjoint. In general, the range of counterfactual situations in which an insensitive belief would be held though false overlaps the range of counterfactual situations in which safety requires holding a belief to ensure its truth. As a result, insensitive beliefs are unsafe.

Counterfactual situations in which my belief that my car is where I left it is false include cases of theft and towing. I continue to hold this belief in such cases, as well as in counterfactual cases in which the belief remains true. The supposition that I continue to hold the belief does not, by itself, make the latter cases any more proximous than the former. I hold the belief in a world in which a thief was deterred from stealing my car only by an unlikely distraction. Why, then, is my belief that my car is where I left it any more safe than sensitive?

We can stipulate that the situation is such that there is no thief, or that he has no interest in my car, or that cars are not being towed. Then the belief is safe, but only because we have pre-empted counterfactual situations that falsify it. No longer can we declare it insensitive. We can weaken the stipulations to give the belief a chance of being false, but make the chance slight—lots of cars, few thieves. Then there is an enormous range of counterfactual situations that preserve the belief and its truth, but these are no more proximous to the presumed actual situation than are (the fewer) situations that preserve the belief while falsifying it. The belief is no longer safe.

The example now begins to approximate the case of lottery beliefs, with the crucial difference, however, that there must be a false lottery belief but there need not be any false car belief. (The probability of lottery beliefs is logical, whereas the probability of car beliefs is statistical.) This difference enables the condition that one's car is stolen to count as abnormal, in my sense—that it has not been stolen is

---

[15] Sosa (1999) claims that safety fares better than sensitivity as a condition for knowledge, because competition with the skeptic obviates safety's proximity restrictions. I don't see how this can be. It seems to me that safety circumvents the proximity restrictions that protect knowledge from skepticism only via the right presuppositions. Depending on what one is prepared to assume as to how things are, safety may be as readily violated as sensitivity. My belief that I have hands is no more safe than sensitive with skeptical scenarios in the running. The beliefs whose positive epistemic status proximity restrictions are supposed to protect against skepticism violate safety if, as things are, the skeptical possibilities that defeat them are proximous. Neither sensitivity nor safety is achieved without the assumption that the actual world is not one to which skeptical alternatives are proximous. But this assumption is impermissibly question-begging with skepticism at issue.

a natural presupposition of reliance on memory to fix its location, whereas the condition that one's ticket wins is not abnormal but merely improbable. So justification differs in these cases (a point anticipated in Chapter 6 and to be developed below). As per Chapter 1, I do not believe that people have lottery beliefs, but I will consider the example anyway for those who do believe this.

The belief that one's ticket will lose is insensitive, because the supposition that one's ticket wins makes no difference to the probabilistic considerations that support this belief. But neither is the belief safe. The counterfactual situation in which it is false has the same probability as any other counterfactual situation in which the belief is sustained. The fact that there are so many more counterfactual situations that preserve the belief's truth is no reason to omit the one that falsifies the belief from the range of situations that preserve the belief. It cannot possibly be any less proximous than they are.

One cannot argue that the situation in which the winning ticket is other than one's own and than the ticket that actually wins is more probable than the situation in which one's own ticket wins. For there is no such thing as *the* situation in which the winning ticket is other than one's own and than the ticket that actually wins. Many situations satisfy this description. Similarly, in the car case, any number of distinct counterfactual situations are such that in them mine is not the stolen car. It does not identify a more proximous counterfactual situation simply to stipulate that some other car is stolen. In cases where safety was supposed to deliver, and sensitivity to withhold, knowledge or justification, we find repeatedly that safety and sensitivity fair alike.

8.6 Luck Versus Knowledge

In my theory, no specification of ranges of counterfactual situations is necessary. As explained in Chapter 7, the truth-values of my subjunctives are determined by the facts of the situation to which they apply (including the justifiedly believable fact that no skeptical scenario obtains), if they are determined at all. And I do not see that safety's restriction on quantification over conditions, what I have called the second step its advocates take, improves upon the original requirement that one not believe P if P is false under any conditions. Even restricted, the requirement is too strong to impose on knowledge. We know lots of things that we not only could, conceivably, under remote conditions, have been wrong about, but also that we would have been wrong about had things gone only a bit differently. It is simply false that knowledge is reserved for the epistemically insured. Sometimes one knows by luck.

After a bitter battle for custody, a child's victorious guardian, unbeknownst to the child and unanticipated by the unsuccessful disputants, takes legal action, out of spite, to change the child's name. As luck would have it, a document is misfiled and the change does not go through. Does the child not know his name because, had a clerk not erred, it would have been different? I say the child knows his name. I know where my car is parked even though realistic scenarios in which I am wrong do not affect my belief. My car might have been stolen. They might have towed my car to make space for a wealthy alumnus. Such a complication is not normal; it is characteristic of situations in which one believes that one's car is where one remembers parking it that one's car has not been stolen or towed. The normalcy of such complications would disrupt ordinary reliance on memory, endangering justification, and thereby knowledge if knowledge requires justification. But as a possibility, it is not far-fetched or remote, not of a piece with skeptical scenarios. It does not violate a standard of proximity or similarity or relevance; it does not violate any standard invented to interpret the accessibility relation for possible worlds. It is just abnormal.[16]

So I conclude that safety, and (according to my argument) sensitivity along with it, are not required for knowledge. Of course, this result does not affect the requirement of counterfactual sensitivity to falsity that I impose on reliable belief, even if justification requires reliability and knowledge requires justification. For reliable belief need not be sensitive to falsity under abnormal conditions. The conditions under which safety and sensitivity are violated without detriment to knowledge are abnormal.

The examples show how luck subverts the explication I entertained for knowing. What if the change of name had gone through, becoming official, but then the new guardian experienced a change of heart and had the original name reinstated? Then would the child know? I remember where I left my car and have a true belief that the car is there. I know nothing of the fact that, in the meantime, the police have misidentified it as a vehicle reported stolen, and have towed it away. Recognizing their error, they then returned it to the place I believe it to occupy. They did this just moments ago; the car could easily have been missing (still) when I went to get it. As I go for the car, do I know where it is?

I have a justified belief and I believe justifiedly that the car is there, and I am right. But in this case I do not know (just as in the elaborated bank case DeRose did not know). It is only by luck, the fortuitous and timely discovery of the error and the remarkable promptness of restitution, that I am right. Had the name change become official, it would be luck that the child is right; the guardian's vindictiveness could have endured. Both the child and I were wrong for a time, and were restored to accuracy through no agency of our own. Had none of this drama occurred, as normally it does not, then we would have known despite the element of luck in its nonoccurrence.[17]

---

[16] Of course, one could adopt my notion of normality to identify the accessible worlds. I raised this possibility in Chapter 3.

[17] As I am not advancing a theory of knowing, it is not my argumentative burden to defend a closure principle for knowledge. But since I have, informally, endorsed such a principle, it is worth pointing out that these examples do not violate epistemic closure. I cannot, by inference from what I ordinarily take myself to know, come to know that the dramas that I say would defeat ordinary knowledge have not occurred. But what I take myself to know does not entail that they have not occurred, for these dramas are constructed to preserve the truth of my ordinary belief. If I cannot know that the police have not removed and then returned my car, epistemic closure prevents me

There is (I have heard) an intuition that the mere truth of a belief cannot, in and of itself, make the difference between knowing and not knowing. I think this intuition is almost wrong. It is wrong in spirit, because what more knowing depends on can be a purely external condition, of which the knower is fortunately ignorant. The child who knows that legal action has been taken to change his name, but not that the clerk erred, does not know his name. The child kept in ignorance does know, despite the falsity of his belief in a world so proximous that only luck prevents him from occupying it.

8.7   Luck Versus Reliability

A different response to fortuitous protection against defeaters is to wonder whether watching CNN is a reliable method after all. Having sorted out how luck affects knowledge and justification, consider its effect on the underlying condition of reliability. The best media resource sometimes errs. Are these times necessarily abnormal, as reliability requires, or is error, though infrequent, built into the way the system works? When CNN is wrong, are conditions essentially different from when it is right?

The case is similar to that of the encyclopedia, in which I contended that it is correct to attribute error to abnormalcy. The point I emphasize here is that it is quite possible for CNN *never* to misinform. This may be unlikely, because it is unlikely for conditions never to be abnormal. But the case is unlike the lottery, for example, in which one is bound to believe falsely if one's method is probabilistic. As emphasized in Chapter 5, probabilities in the lottery are logical probabilities, in the sense of being *a priori* relative to the defining terms of the exchange. These terms require that some improbable event occur. A merely statistical probability of error projected from past experience does not guarantee (further) error.

In fact, (bracketing my general reservations about the determinacy of statistical probabilities) one could defend low statistical probability of error as a truth-conducive standard of justification. By this standard it is strictly possible to get truth while avoiding error. My reason for rejecting this standard is not that it renders false belief inevitable. Rather, given CJC, whatever threshold one fixes for justification, a probabilistic standard has the consequence that unjustified beliefs will be more probable than justified beliefs. The low probability of a justified belief is tolerable as the result of the application of reliable methods under abnormal conditions. It is tolerable as the result of CJC, provided that justification is independent of probability. It would be intolerable (incoherent) with probability itself the standard of justification. Any probabilistic standard of justification must sacrifice CJC. This is sufficient reason to reject such a standard.

---

from knowing that I know where my car is. But it does not prevent my knowing where my car is. What would pre-empt this first order knowledge is an inability to know that my car is not now somewhere else.

A further reason, introduced in Chapter 3, is that it is unacceptable to interpret justification in a way that makes success at advancing the epistemic goal a mere possibility whose realization is overwhelmingly improbable. Success is something that the use of justificatory methods should entitle us to expect. It must not depend on improbable accident, if pursued by means that conduce to truth.

If believing CNN is justificatory, then it cannot be purely chance for CNN not to be wrong. Mistakes may be likely, but it cannot take an accident to avoid them. Learning of the double, CNN should have said that the captive's identity is now in question, not that the terrorist remains free. There are standards for reporting that CNN, abnormally, violated. They said that Gore won Florida, when they should have said that the majority of voters intended to vote for him. If the pertinence of this distinction is not abnormal, democracy is useless.

Strictly speaking, of course, it would have been too strong to say even that Gore had majority support. This conclusion came from exit polling, and polls can certainly be wrong under normal conditions; they are wrong, on average, 5% of the time, where the standard of accuracy determining the margins of error is based on two standard deviations. Any report of polling, any conclusion about a population based on sampling, must be read as including both margins of error and a (logical, not statistical) probability of inaccuracy beyond these margins. So CNN should really have said that there is x probability that Gore's support exceeded Bush's by at least y percent, for specific x and y determined by their polling numbers. This statement would be infallible under normal conditions, for it assumes only that respondents did not lie, did not misremember their intentions, were not robots infiltrated by FOX to sabotage CNN, and so forth.

While it is unusual for popular media to be so precise, it is not unusual for them to issue caveats as to margins of error, and even as to the probabilistic flexibility of the conclusion. Common locutions like "We project that—" and "We predict that—" concede the probabilistic status of the forecast and possibilities for inaccuracy of the data in the sample. Of course, such statements are unqualifiedly true under normal conditions. I grant that often such complications must be read in if a media report is to meet the standard of reliability. But I submit that reading them in is fair; it does not change the intended content of the report. Reading them in, I trust CNN not because of its high statistical probability of accuracy, but because of the intelligence and professionalism of its reporters and its reputation for thoroughness of research. And I do not judge it unreliable because these grounds fall short of perfection.[18]

One might wonder, however, whether these grounds warrant imputing so strong a condition as I take reliability to be. What if a method yields falsehoods under normal conditions, but is self-correcting, such that eventually (ultimately?) the truth will out? Suppose it is not abnormal for CNN to misreport, but it is abnormal for its

---

[18] By contrast, the troubles at the *New York Times*, regarding the Jayson Blair affair, go right to the ideology of its editorial practice. Can I uphold my confidence in CNN in the face of systemic malfeasance in so venerable an institution as the *New York Times*? I can. That institutions of CNN's type have gone bad is second-order evidence against reliance on CNN. I have disputed the priority of second-order evidence in Chapter 6.

mistakes to go uncorrected. We might even speculate that the only effective route to knowledge in some domain leads via falsity or deception, which must be tolerated as a temporary evil for the sake of our epistemic ends. False scientific theories can be epistemically progressive. Should I weaken reliability to require only that continued application of a method offer compensation for its errors?

I think not. I have already accommodated some of the intuitiveness of this proposal in allowing that reliable methods can be applied with varying skill or diligence, producing correctable error. I do not go beyond this concession to accept false belief as a reliable method's normal output. Nor have I reason to stipulate that deception is anywhere the necessary means to an epistemic end. Progressive scientific theories are partially true; believing them is not deception. The epistemic goal, as I understand it, is not just to believe truly on balance in the long run, or at the end of inquiry. We seek positive epistemic status along the way. To pursue the goal successfully is to add truth while avoiding falsehood, and justificatory methods are methods that enable us to do this.

Chapter 9
Intellectual Virtue

9.1 Coherence Requirements

My theory does not require for one's justification of a belief that one possess a general capacity or faculty for discriminating truths from falsehoods across a range of propositions to which the belief belongs. Such a faculty is a kind of intellectual virtue. It abstracts away from particular cognitive states or acts to an abiding feature of intellectual life, much as moral virtue abstracts away from particular right acts to stable traits of moral character. This chapter explains why some philosophers require intellectual virtue for justification, and why it is a mistake to do so.

Reliable methods are general methods and must be reapplicable on different occasions of belief-formation. But the beliefs a reliable method is used to form need have nothing more in common to group them as the objects of a single faculty or competence than do the entries in an encyclopedia. I do not even rule out that one can justifiedly believe P without believing anything else, as Descartes (supposedly) believed himself to be thinking while in doubt as to all else. But if this is extreme,[1] certainly justification can be limited and fleeting. One does not have to be very good at the epistemic enterprise to be credited with getting *something* right.

In fact, it amounts to an objectionable elitism to withhold justification altogether where intellectual life fails to flourish in a general way. Occasional justification amid a morass of superstition and ineptitude is commonplace. How else are the epistemically underprivileged to progress? One must have some justified beliefs in order to develop intellectual virtue, just as one must do some things right to develop moral virtue. Intellectual virtue is developed through comparing justified beliefs

---

[1] It is difficult to imagine being in a position to *attribute* just one belief. Even Descartes, upon reflection, professed further beliefs as to what he was thinking about and what form his thinking took. Though confined, he executed some motion within the space of reasons.

J. Leplin, *A Theory of Epistemic Justification*, Philosophical Studies Series 112, DOI 10.1007/978-1-4020-9567-2_9, © Springer Science+Business Media B.V. 2009

and with others and recognizing the differences.[2] So justification cannot depend on possession of virtue.[3]

I reject the competing coherentist picture of virtue, and with it justification, as cumulative conditions that accrue to mental functioning gradually, as behavioral dispositions evolve into belief systems that develop a richness and sophistication of mutually reinforcing relations. On the coherentist picture, beliefs are justified only derivatively from their participation in a sufficiently evolved system; it is the system as a whole that is the proper object of appraisal.

My objections to this alternative are familiar. I do not understand talk of systems as a whole, and especially not talk of their appraisal. Such talk would have to be part of the system, with rampant relativism the consequence. I do not think that the gradual, emergentist picture translates easily from the evolutionary context that fostered it to an understanding of the epistemic progress of individuals. It seems to me a plain fact that people typically have both justified and unjustified beliefs, and that the difference is not a matter of any states or capacities that characterize the believer generally. Rather, the difference is highly sensitive to content and method. The coexistence of radically divergent epistemic states within belief systems is the norm; there is no reason to expect epistemic strengths or weaknesses to be uniform. That some beliefs are justified is data for me; general conditions of coherence are, by contrast, abstract and inferential. The idea that justification awaits a broadly coherent and grounded intellectual perspective that evaluates and situates propositions according to general principles consistently executed strikes me as unreasonably and unrealistically exigent. Why can't things go well epistemically just as far as P, and collapse after that?

One answer is that even if justification is not identifiable with general coherence, it is nevertheless constrained by requirements of coherence that such a possibility would violate. Consider again cases of external manipulation. It is not automatic that we can retreat to an internal level at which justification is unaffected. Suppose that the mad scientist misaligns the beliefs and impressions that he feeds me. When I see red I believe green. Even if my beliefs work together, they may be systematically incongruous with my experience. Suppose the selection of beliefs to induce is random. Even if they happen on a particular occasion to coalesce into a coherent system, there is a respect in which coherence is violated, for the system

---

[2] Then must one *also* have unjustified beliefs as a condition of developing virtue? I am inclined to think so, although the condition is so easily satisfied as to be untestable. The analogy to morality suggests to me that some measure of irreverence and disobedience is essential to moral development. Someone who never breaks the rules never gets beyond them to the possession of moral character.

[3] But justification can depend on the ability to apply a method of belief-formation. If the very ability to form a belief counts as a virtue, then justification can require virtue. But the very ability to form a belief is not the sort of virtue that has to be developed. It is not a general faculty or competence. It could be possessed just once or just fleetingly, as when the blind become momentarily sighted.

as a whole does not develop in a coherent way.[4] Suppose the scientist gives me contradictory beliefs but not their conjunctions, nor the capacity to scrutinize and notice inconsistencies, so that each belief appears justified viewed individually. In such cases it is tempting to reject the possibility that the victim is justified in *any* of his beliefs. The general internal mess that the scientist has made might seem to pre-empt all justification, and possession of some general intellectual virtue might seem the missing condition.

Although there is certainly a problem of coherence in these suppositions, I want to suggest that it is a problem not about justification but about belief itself. I begin by noting that coherence has not been at issue in examples based upon external manipulation. The argument against reliabilism from external manipulation is supposed to be that beliefs can be justified despite being unreliably produced if the subject has no reason to distrust the reliability of their source. An unreliable external source is consistent with the coherence and apparent veridicality of the beliefs induced. There is also the converse argument that beliefs are unjustified despite being reliably produced if the subject has no reason to trust, or has reason to distrust, the reliability of their source. Perhaps such distrust can be located in problems of coherence. But that the beliefs produced may be random or discordant in some way, with one another or with experience, is a different objection, which has nothing essential to do with an external manipulative source. Is someone unmanipulated but mentally unbalanced, or subject to unpredictable periods of hallucination, incapable of justified belief? So then are we all, for we dream.

Perhaps the virtue theorist means to be making a *new* point against reliability, that coherence is necessary for justification and the reliability of a source of belief, whether internal or external, manipulative or an exercise of the will, does not guarantee coherence. Perhaps someone reliably believes thermometer readings, but his temperature beliefs are completely discordant with all his other beliefs, including beliefs about the weather, about sensations of hot and cold, about the effects of heat on physical objects. Then his reliable temperature beliefs are unjustified. What if the *truth* is not coherent; that is, what if the natural world is too chaotic to support induction and reliable methods yield a haphazard jumble of disconnected beliefs? If the world were like this, our beliefs about it, even if true, could not be justified. External manipulation matters to this objection only as the extreme case in which there are no constraints on what comes to be believed.

But what is hard to make sense of in scenarios, whether internal or external, that produce radical incoherence is not that certain beliefs are held justifiedly but that the subject's state is actually one of investing credence, of epistemic commitment, to begin with. Coherence is misplaced as a standard for justification. It is, in the

---

[4] Here I am trying to understand Ernest Sosa's position in "Reliabilism and Intellectual Virtue" in (1991). Sosa thinks that because of such possibilities, justification must depend on the possession of intellectual virtue. I will presently be engaging Sosa directly.

first instance, a standard for believing.[5] Presented with quixotic randomness, mutual inconsistency, or disconnection from reality in a subject's pronouncements, one is unable to say what if anything the subject believes. A mind in a jumble cannot be pinned down, whereas the attribution of belief is necessarily a basis of expectations for future thought and behavior. That such expectations are systematically belied disconfirms one's belief attributions.

Moreover, if it is a condition upon doxastic interpretation that the subject's beliefs turn out substantially true in the interpreter's system, and this condition cannot be met, then the subject, as interpreted, can have no beliefs. And it seems to me that the subject himself cannot say what if anything he believes, cannot consistently interpret his own state as one of believing, if he systematically violates conditions that he would require for the attribution of beliefs to others. His understanding of what it is to believe must consist in such conditions.[6]

9.2 Induced Belief?

To suppose that the mad scientist can make one believe just anything, however incongruous, is to ignore coherentist constraints on the nature of belief that hold independently of any question of justification. I have suggested that what the scientist imposes are, at most, impressions, images, sensations, thoughts; whether these instigate belief must depend upon the subject's internal processing.[7] A thought is not a belief unless endorsed by the believer; that it be made to appear to the believer that he endorses it is insufficient. Perhaps it can be made to seem to the victim that he has these beliefs, but if so this appearance is corrigible. A person confuses Austria with Australia and thinks he believes that black swans were discovered in

---

[5] Interestingly, the arch (though erstwhile) coherentist Laurence Bonjour appears to concede this priority. "It would be reasonable," he writes in parentheses, "to regard a reasonable degree of stability as a necessary condition for even speaking of a single ongoing system of beliefs" (1985, p. 170). But then he declines so to regard it, opting instead to make stability a condition for justification. (I am unclear what the qualification "single ongoing system" is doing in this fleeting concession). Bonjour's more recent writings depart from coherentism for internalist reasons (see his 2003). Internalism requires the justified believer to have beliefs as to the contents and coherence of his system of beliefs, and the recent Bonjour worries that these required beliefs cannot be justified on coherentist grounds without circularity. Of course, this worry arises only if *all* justification is internal justification.

[6] How much coherence does this reasoning require? Could coherence be fleeting and yet beliefs endure? Does the subject continue to self-attribute beliefs that no longer cohere? I suppose that beliefs could endure in the face of discordant experience, but an incoherence among beliefs will require some change in content that subverts their status as continuing beliefs.

[7] It is possible to drive coherentist requirements deeper: even impressions and images must cohere, must be conceptualized in some stable way. Intentionality, it is said, is the mark of the mental. Well, I think it is the mark of some of the mental, but I see no reason to place imagery beyond the powers of unconstrained manipulation. Those who do will grant my argument against virtue theory all the more readily.

Austria. I see no need to agree with him that this is what he believes. A religious biologist thinks he believes that evolution by natural selection is compatible with divine creation, but really he knows better. If knowing requires believing, then he is wrong about what he believes.

Saying this seems to attribute to the believer a belief that he has a belief which in fact he lacks, or a belief that he lacks a belief which in fact he has; and perhaps it is within the capacity of the grand manipulator to induce second-order beliefs, though not first-order ones. Perhaps the impression that one endorses a thought is sufficient to compel one's endorsement of this impression. For we do have beliefs about what we believe, and these beliefs must be able to originate in a way that does not guarantee their veracity.[8] I am untroubled by this prospect. I doubt, for example, that anyone actually believes in a Christian, personal god, although lots of people believe that they do. The belief that one does is consonant with one's other beliefs, one's behavior, one's experience, and with principles of rationality and coherence; the first-order belief manifestly is not. Where supposed first-order beliefs violate coherentist constraints, the second-order beliefs we are inclined to attribute satisfy them.

On a material theory of mind, beliefs are physical states. Can't the scientist simply implant these states? What is to stop him? If it is the agency of the believer that is missing, why cannot this agency itself be implanted? What is this agency, anyway? It cannot be merely an image, impression, or thought, for these could be illusive; one could have an illusion of agency. And, on pain of regress, it is not a further belief. I think of this agency as a kind of action, despite the fact that many beliefs are dispositional. Maybe dispositional beliefs lying in storage must have received an original endorsement. Dispositions to believe are not beliefs unless and until entertained, at least fleetingly. The correct conclusion of this line of argument, it seems to me, is that a developed material theory of mind must reconceive beliefs as complex relational properties, or, perhaps, eliminate them altogether, as molecular biology has either reconceived or eliminated the gene.[9] A belief cannot be treated as an isolable physical state introducible or eliminable independently of other states. For it makes no sense to attribute a belief to a subject without regard to what else he believes, says, and does.

The scenario of the mad scientist, that supposedly yields justification without reliability, is in one respect more challenging to externalism than the converse scenario of the clairvoyant, that supposedly yields reliability without justification. What the

---

[8] Certainly a (trick) belief like the belief that one does not believe something falsely must be able to originate in such a way, but I mean beliefs as to the semantic content of one's beliefs. For trickery, see Chapter 10.

[9] These cases are difficult. Philosophers of physics continue to disagree as to whether the electromagnetic ether does not exist or is an electromagnetic field. (My view is that genes do exist but the ether does not.) As for eliminative materialism, I myself find the position (virtually) incoherent, but I know that eliminative materialists think they have a ready explanation for that. I suppose that if the very concept of belief goes the way of witchcraft, then much epistemology, including conceptual studies like this one, will be relegated to the history of ideas.

scientist induces are impressions, images, thoughts, sensations. These are supposed to form a coherent total picture, as coherent as one's veridical picture in normal experience, for the skeptical point is that the actual picture could be nonveridically imposed. Given this picture, one forms beliefs; the picture authorizes and sustains beliefs. It is not the scientist who induces belief, but the subject himself. What the scientist does is to set the subject up, to trick him into investing credence. But the subject *must be tricked*; the impressions must be such as not to arouse suspicion as to their veridicality. This requires coherence, reinforcement, continuity. Though not generally voluntary, the act of investing credence is the subject's. If the subject happens to be a skeptic, the scientist loses.

By contrast, clairvoyance is supposed to instill beliefs directly. The subject just finds himself believing things, with no apparent basis in experience or prior belief. It is not just thoughts or impressions that pop into his mind, but convictions. Otherwise, there is no challenge to reliability as a source of justification. But this is very strange. The subject has not made up his mind, has not entered into an epistemic commitment, has not *decided* anything. For the subject has not *done* anything; he just comes into possession of beliefs, as one comes, helplessly, into possession of fruitcake at Christmas. He notices that he has beliefs as one might notice a fragment of walnut shell in the cake. He cannot attribute these beliefs to perception, memory, immediate inference, or any other relatively passive mode of belief-formation understandable to him.

How can we take this seriously without expecting these beliefs to be suspended immediately upon their recognition, with no more difficulty, nor less alacrity, than discarding the cake? Does the subject find himself *unable* to suspend a judgment that he believes baseless or refuted? What incapacitates him? Unless massively confused, he cannot sense a responsibility to honor what arrives unsolicited. An impression or thought, perhaps a *feeling* of credulousness, we may imagine indelible, but not a conviction. And if the belief does not survive even cursory scrutiny, surely it is incorrect to attribute it in the first place.

Clairvoyance in fiction consists of visions of events inaccessible in the subject's experience, events distant in space or time. The subject might come to trust these visions by induction on their (subsequently experientially corroborated) actualization, but this requires a further judgment by the subject. He must do something, and not merely be done to. Only in philosophy does clairvoyance supply beliefs, and for good reason: philosophical clairvoyance is too unbelievable to work as fiction.

Laurence Bonjour (1985, Chapter 3) originally deployed clairvoyance cases as counterexamples to externalist justification, to reliabilism in particular. But despite what Bonjour says, it is unclear that his cases *do* challenge reliabilism. For although Bonjour insists that the subject's presumptively unjustified beliefs result from the operation of a reliable clairvoyant cognitive power, Bonjour requires the subject to *believe* that he possesses this power and questions whether the subject can maintain his clairvoyant beliefs *without* this conviction. Now, is the belief that one is clairvoyant itself a belief that clairvoyance delivers? Is *this* belief clairvoyant? If so, there can be no reason, *contra* Bonjour, to require the subject to have this second-order belief *in addition* to his specific, first-order clairvoyant beliefs. By hypothesis, clairvoyance *all by itself* is reliably supplying beliefs that need not, on that account, be justified. And if the belief that one is clairvoyant is *not* itself clairvoyant, then any dependence upon this belief of one's (purportedly) clairvoyant beliefs usurps the status of clairvoyance as a presumptively reliable method of forming beliefs. The subject's method of believing is not, then, clairvoyance, but rather (something like) the subsumption of clairvoyantly supplied imagery, or whatever, under the general rubric of a nonclairvoyant disposition to credence.

Bonjour does not consider whether the belief that one is clairvoyant is clairvoyant. If he did, I do not think he could maintain clairvoyance as a counterexample to reliabilist justification. He would have to concede a role for the subject's own agency in belief-formation.

A comparison of epistemic conviction to moral obligation is apt. Only the agent can obligate himself morally.[10] He can be maneuvered into obligating himself, but his obligation cannot be maneuvered into him. He can be maneuvered into thinking he has an obligation, but not into having one. In the same way, belief requires the agency of the believer. You can no more make him believe by imposing a feeling of credulity than you can make him indebted by imposing a feeling of guilt. Compelling visions that one cannot ignore, possibly; but why call these beliefs? Make him feel, think, experience what you like, you cannot make him believe without his cooperation. Thoughts that come at random, that clash with experience, that violate expectation, that fail standards of coherence will not get this cooperation. They will not become beliefs. As justification is at issue only for beliefs, it is not at issue where massive failures of coherence make the attribution of belief implausible and unnecessary.

9.3 Virtue Versus Reliability

I submit that coherentist constraints on belief do justice to the intuitions that promote virtue theory. Whether or not I am right about this is for those possessed of these intuitions to decide. Virtue theory is unacceptable in any case, from the perspective of reliabilism. This is not recognized, because virtue theory is typically portrayed as a *gloss* on reliability, as if virtue theorists are a species of reliabilists.[11] But the gloss is at once too demanding and too permissive for reliabilism to wear it.

The idea of intellectual virtuousness in general is, of course, vague. Rather than seek full justice for the virtue theorist's intuitions, let us stick to the theory's core idea that justification is a collective condition that is not achievable for beliefs taken one at a time, and to the core thesis, due to Ernest Sosa, that one is justified in believing *P* just in case one possesses a special competence for judging the truth-values

---

[10] Or so it seems to me. If you disagree, try legal obligation.

[11] Ernest Sosa assumes this connection throughout his writings.

of propositions across a range to which P belongs.[12] Then instead of reliability in the formation of individual beliefs, we have reliability with respect to a class of propositions. One's ability to sort these accurately by truth-value is supposed to justify all of one's beliefs within this range.[13]

Now I have two questions. First, does the justification of P depend on the full exercise of one's evaluative ability across P's entire range? That is, must one have made up one's mind about everything in this range, or might some of its members fail to be entertained, or be entertained without resulting in belief or disbelief? Sosa *seems* to require that an intellectual virtue be fully exercised to justify at all, but I do not understand the motivation. I assume that such a requirement is unreasonable and unrealistic for any but a trivially circumscribed range.

Secondly, must all the beliefs one *does* have within P's range originate in the same way as P? For P to be justified, is it necessary that other beliefs in P's range be formed as P was? Granting that there is a single faculty reliable throughout the range, must it be *this* faculty alone that one uses in coming to believe what one does believe within the range? Or, if not alone, must exercise of this faculty at least accompany whatever else one does in forming these beliefs? I am not sure whether Sosa intends this restriction. His official definition of what it is to believe "out of" an intellectual virtue looks satisfiable without the involvement of this virtue in the actual formation of the belief. In particular, the "inner nature" in his conditions for possessing intellectual virtue need only be a disposition to use reliable methods.[14] A diversity of methods might be used.

Whether intended or not, for any interesting range such a restriction is untenable. Nero Wolfe will not suffer the exertion of inclining his head to bring the clock on the wall within his field of vision; he asks Archie Goodwin the time.[15] His resulting belief is justified (by testimony), but what is to exclude it from the range of beliefs for which his perceptual faculty is virtuous? The ranges of beliefs for which different

---

[12] This is Ernest Sosa's thesis in "Reliabilism and Intellectual Virtue" in (1991). The complications and qualifications he introduces in "Intellectual Virtue in Perspective" in (1991) and in (1993) only reinforce what I want to say. Sosa assumes a frequentist interpretation of reliability, but I take this to be incidental to the points at issue. If it isn't, so much the worse for virtue theory, as I have shown that (given the epistemic goal) reliability cannot be so interpreted.

[13] I think "range" is better than "class". A class is defined by its membership, and this is too exacting for the virtue theorist's purposes. Sosa talks of a "field" of propositions. This terminology suggests additional structure not applicable in context.

[14] According to Sosa, one possesses an intellectual virtue with respect to a field of propositions if and only if one has an inner nature such that one is very likely to believe correctly within this field and unlikely to believe incorrectly within the complement of this field with respect to an allowable broader field. To believe a proposition "out of" intellectual virtue is to believe the proposition while possessing such a virtue with respect to a field to which the proposition belongs. For details, see Sosa, 1991, pp. 286–287. This definition does not explicitly restrict the genesis of the specific belief. In other formulations, Sosa speaks of a belief's being made true by an exercise of one's virtue, and being true because of one's intellectual competence. I find these latter accounts virtually unintelligible. What makes a belief true is the cooperation of the world.

[15] These are characters in a series of detective novels by Rex Stout.

faculties are virtuous intersect. Within their intersection, the virtue theorist ought not to restrict the choice of faculty on which justification depends. It would better serve the purpose of virtue theory to come down hard on one's competence with respect to a range of propositions, and then, with this competence established, to be lenient as to the justificatory sources of beliefs in individual propositions that the range contains.

An apparent advantage of this needed largesse is to accommodate the justification of beliefs that are not formed or sustained by any identifiable, reliable process. Maybe there is a reliable intuition that does not operate through any reconstructible sequence of steps, that has no structure in thought, but delivers a verdict immediately upon a proposition's being entertained. This is rationalism, and examples might be recognition of logical or mathematical truths, or the *cogito*. The former might be treated separately, reliabilism being a theory of the justification of contingent beliefs.[16] But the *cogito* is contingent, and room should be made for it. If it is a belief that one cannot have unjustifiedly, then it cannot matter to its justification—let us say, more generally, to its positive epistemic status—how one gets it.

But surely it does matter. Suppose someone is mistaken about who he is. He believes there is a conspiracy behind the death of the President. There had to be an assassin, and he believes himself to be this unknown assailant. Of course, as this is a conspiracy theory, there is no evidence. And in fact, the President died in his sleep. Does the conspiracy theorist believe unjustifiedly in his own existence?

The reply I expect is that in believing oneself to be the assassin one already believes oneself to exist, and this belief must be justified. In having *any* belief about who one is, whether or not the person one takes oneself to be exists, one necessarily believes oneself to exist. I can concede this. My question is what ensures this belief's justification? Why can't one believe that one exists *in virtue* of believing oneself to be the (nonexistent) assassin? Descartes argued not that it was impossible for him to believe in his own existence without being justified, but that it was impossible for him to believe this without its being *true*. He thereby *constructed* a justification. If no justification is constructed, why does there have to be one?

Of course, Descartes had not just a rational intuition, but also a method (intentionally deployed, by the way) of investing credence: believe that which one finds oneself, upon reflection (including application of the intuition), incapable of doubting. Unfortunately, Descartes professed himself incapable of doubting, among other crazy things, that there is at least as much formal reality in the cause of an idea as there is objective reality in the idea it causes. He could even doubt that $2 + 3 = 5$ (the one thing he got right), but he couldn't doubt his causal principle. So this is not a very reliable method, for him anyway. And yet his belief in his own existence was justified. What justified it was the recognition that it could not be believed if false. If it is possible for one to believe oneself to exist *without* this recognition (which, after all, is supposed to *originate* with Descartes), then how one does come to or

---

[16] I will further address the scope of reliabilism in Chapters 10 and 11.

sustain the belief matters to one's justifiedness. The *cogito* does not, then, require the largesse that I have argued that virtue theory needs to accord justification.

This largesse is implausibly permissive. The possession of a competence across P's range, even a competence that one rationally self-ascribes, does not constrain the actual nature of P's formation. Suppose that the possessor of intellectual virtue generally (frequently or in normal conditions, as you prefer) relies on this virtue within P's range and so generally believes correctly. He satisfies virtue-theoretic requirements for justification across the range. P itself he could believe for any crazy nonjustificatory reason. Should the virtue theorist then revert to requiring that P be believed virtuously, as a result of applying to it the competence one possesses over its range? This excludes alternative competences or methods that are perfectly justificatory.

At most the requirement should be to believe P by application of *some* competence that one possesses across *some* range to which P belongs. But now justification has refocused on P individually. It is P itself rather than any particular range that decides what modes of belief-formation are justificatory. If P rather than a range is the proper object of justification, what compels one to apply the relevant competence to anything else? Then P alone is justified; we are back to individually justified beliefs that may not cohere with what else one believes or experiences, and the collectivist character of the theory is lost. Will the virtue theorist allow P to be formed by any method that yields it reliably, whether or not this method is justificatory across a range to which P belongs? Then it is unclear what other than reliability virtue amounts to. What do we need virtue theory for?

9.4 The Scope of One's Competence

What is really going on, it seems to me, in the shift from reliability to virtue is a change of focus from how a belief is formed to the kind of belief it is. To be justified, a belief must be of a kind that the believer is and recognizes himself to be qualified to judge. The virtuous believer has a perspective on his range of competence. What matters is not how the particular belief arises, but that the believer's qualifications extend across the entire kind. This requirement prevents the justification of beliefs issuing from a normally inerrant cognitive faculty, if the subject cannot distinguish these from others of his beliefs that originate unreliably. The subject must grasp the scope of his competence. In an example of Sosa's (1991, 1993), the subject's vision is flawless but he cannot tell (cannot sort) the beliefs his vision yields from the deliverances of hallucinogenic intervention. Sosa contends that although vision is reliable, it cannot be justificatory in the presence of systematic unreliable production of beliefs of the same general kind. His prescription is the additional requirement of an intellectual virtue with respect to beliefs of this kind. The subject must get these beliefs right, across the board, which he does not if he hallucinates.

From the perspective of my reliability theory, the example is underdescribed. The correct diagnosis as to justification depends on the nature of the connection between the subject's faculty of vision and his vulnerability to hallucination. If judging by sight causes him to hallucinate some proportion of the time, if it creates or increases his vulnerability to experiencing imagery that he mistakes for veridical seeing, then the faculty of vision is defective and unreliable as a source of belief. If a neurological disorder or external manipulator induces hallucinations as a function of the subject's exercise of sight, then although not itself culpable (so to speak) the faculty of vision is nevertheless rendered unreliable by this interference. Under normal conditions, the intention to judge by vision results in false beliefs. We therefore have the result Sosa wants in this case, that beliefs based on vision are unjustified, without imposing any virtue-theoretic requirement beyond (my version of) reliability for justification.

However, if there is *no* connection between the faculty of vision and the episodes of hallucination, if it is just that hallucinations, when they do occur, produce what the subject cannot distinguish from sight, then I do not share with Sosa the intuition that the subject's vision cannot be justificatory. The subject gets justified beliefs by seeing when that is what he is doing, and unjustified beliefs that he is justified in believing when he takes himself to be seeing but is hallucinating instead. My distinction between justified believing and justified belief makes perfectly adequate sense of what is going on; there is no need to deny justification altogether. Sosa fears that if the subject's beliefs are justified when he sees, then the subject's beliefs will have to remain justified when he faultlessly but mistakenly takes himself to see. Not to worry; it suffices for the subject himself to be justified in the latter case.

Of course, if the subject *finds out* that he was hallucinating when he took himself to be seeing, if he determines that his purportedly sight-induced beliefs regularly turn out erroneous, then he loses confidence in vision as a basis for belief. But I do not think this means that he loses justification for these beliefs, and I do not infer, with Sosa, that justification requires a reflective confidence in one's cognitive functions.[17] Rather, this means that the subject *no longer forms* beliefs on the basis of apparent vision at all. He ceases to invest credence under conditions that he has learned not to trust.

Is this psychological speculation? Could the subject not defy me? I find it hard to understand the suggestion. It seems to me that to lose confidence in one's vision *is* to suspend belief based on vision. But if we do imagine, somehow, that the subject continues to believe that things are as he seems to see, despite his uncertainty that he really sees, then my conditions for believing justifiedly are unfulfilled. The subject no longer has good reason to believe his method reliable. There is no need to impose Sosa's stronger requirement of reflective confidence to pre-empt justification.

In the general case, unqualified by consequences of what the subject learns or suspects about the reliability of trusting apparent vision, my theory, in opposition to Sosa, grants the subject justified beliefs from vision. I do not see why the proportion of these beliefs among those the subject *takes* to be justified by vision cannot be small. Kepler had a few great scientific insights amid a morass of superstition and wanton metaphysics, without himself being able (reliably) to tell the difference. The

---

[17] Sosa, "Intellectual Virtue in Perspective", in (1991).

timid can have moments of courage, the confused moments of lucidity. It may only be an external perspective that distinguishes such moments. Why impugn the beliefs that reliable vision delivers, just because a lot of error goes along for the ride?

9.5 Social Epistemology

The reason, for Sosa and many others, is a feel for the social importance of justification. If the subject cannot tell justified from unjustified beliefs, if his credulity cannot discriminate, then he is unreliable as a source of information. In fact, I suspect that this concern is responsible for the failure of most epistemologists to distinguish the justifications of beliefs and believers. If what we value in the subject is his epistemic contribution to society, then the former form of justification does not matter. For it does not affect action or testimony. Justified in believing justified and unjustified beliefs alike, the subject is not positioned to participate in the communal disposition of epistemic responsibility. He cannot do his part.[18]

I am afraid the same dysfunction potentially attends any faculty with social utility. A witness with a personal grievance cannot be trusted; it does not follow that he testifies in error or that he would allow his grievance to influence his testimony. He may be fully trustworthy, although we cannot trust him. An expert carpenter cannot be relied upon if he shows up drunk (an unpredictable) half of the time and, when drunk, cannot tell a good job from a sloppy one. It does not follow that his carpentry is never successful. I think that the social dimension of justification is misrepresented in arguing that the beliefs of the victim of hallucination cannot be epistemically successful. It is we who are not justified in believing him, not he whose beliefs are unjustified. To be justified in trusting someone's beliefs, in taking what he believes to be a source of information, it is insufficient that these beliefs be justified. We must be justified in believing they are justified.[19] The reliability of the victim is a matter independent of the reliability of the victim's beliefs. The former applies to our method of forming beliefs, which employs him. The latter applies to his method, which employs, say, vision.

There is, of course, a broader agenda to the socialization of epistemology. The thesis that epistemic justification is social really means that the condition of being justified cannot arise in isolation, that one's own justification depends upon

---

[18] This diagnosis is a bit sweeping, for there are many ways to serve. Justified in holding beliefs that we are justified in believing unjustified, the subject can serve us as a cautionary model. Perhaps, then, the assimilation of justified belief to justified believing is associated with valuing information over criticism.

[19] I assume for simplicity that if we were justified in believing his beliefs are true, which would suffice for trust, we would thereby be justified in believing they are justified. Of course, the converse does not hold, so (our) justification is insufficient for trust. We could justifiedly believe that his beliefs are justified and also that they are formed under abnormal conditions.

the epistemic condition of others. Reasons, grounds, and evidence are important just because social values like respect, empathy, and civility depend upon sharing them with others. Robinson Crusoe's beliefs need not be justified; they need only be true.

From the perspective of my theory, this line of argument is convoluted. Supposing that it gets the provenance of communal values right, what follows is not that one's own justification is hostage to that of others, but that the health of society is hostage to the possession by individuals of justifications accessible to others. Accessibility requires clarity of understanding; one cannot effectively share what one does not understand. And understanding depends on social interaction. A person's reasons for belief, as for action, are often less evident to himself than to others. But the justification of believing only requires possessing reasons; it does not require understanding them.

9.6 Virtue and Transmission

I suggested that virtue theory is both too demanding and too permissive to be grafted onto a reliabilist epistemology. And more serious to me than the largesse I have criticized is the respect in which virtue theory is onerous. A purpose and benefit of reliability theory, as I understand it, is to guarantee the transmissibility of justification through truth-preserving inference. Virtue theory abrogates this guarantee. For competence across a range to which the beliefs one infers from belong does not guarantee competence across a range to which the belief one infers belongs. Inferential closure fails for virtuousness. Of course, the intended notion of a "range" ("field", says Sosa) of beliefs is vague, possibly to the point of emptiness. If just any set will do, the condition is satisfied trivially. I grant that there is more to the idea than this, but any plausible and nontrivial way I can think of to pin it down carries the unacceptable consequence that inference cannot be justificatory.

Minimally, P's range ought to include the alternatives to P that are contextually viable candidates for credence. For in judging P's truth-value, one judges theirs as well. Beyond this minimum, the idea would be to include some propositions of related semantic content, perhaps some nonexclusive alternatives, such that what one does to judge P is also applicable to them. Already the expansion is elusive, so consider an example.

Jones, whom I know only slightly, tells me he is hungry (let this, that Jones is hungry, be P), under conditions in which this testimony is appropriate and plausible. Of course, I believe him and am justified in doing so. If one could not be justified in one's beliefs about the mundane mental states of others under normal conditions, the social order would collapse (and so it has, where suspicion is routine). Moreover, let it be true that Jones would not normally dissemble on the point of his appetite, so that my belief is justified. I also justifiedly believe, occurrently should a question of how I know arise, that Jones told me he is hungry (Q). Were Q false I would not believe it, so this belief too is justified. From these beliefs I can infer that Jones spoke sincerely (R).

Now, I grant that P and Q belong to broadly (loosely) specifiable ranges within which my justification and that of my beliefs are unproblematic. P's range includes Jones's "self-regarding" propositions, propositions reporting mental states of Jones that I have no ready evidential basis for judging apart from Jones's own testimony; that he is thirsty, tired, impatient (and their negations), and alternatives like his being (not hungry but) inclined to do the social thing, which in this case is eat. Q's range includes other overt acts of Jones; that he sings, whispers, sits, stands, paces—I must leave it to the virtue theorist to decide how far to go with this. All these I can judge just as I judged P and Q.

But, compatibly with this expertise, within R's range I need have no general competence. Evidently R's range contains acts exibiting traits of character—compassion, conscientiousness, disingenuousness, affectation—that I may be unable to judge. Yet I can by inference come to believe R, and this inferred belief is justified. Sincerity I can infer from the truth of what is said, if the matter is one the speaker would not normally misjudge. Criteria for the application of more complicated traits could elude me, even if I am good at self-regarding propositions and overt acts. My *general* competence exhibited with respect to P and Q need not extend to R.

I do not imagine the virtue theorist responding by identifying inference itself as a virtue. I do not imagine him simply tacking closure on to his theory in this way. Inference does not sort true from false propositions within an independently circumscribable range, as uninferred propositions are left undecided. This move abandons the role of the range of competence, which distinguishes virtue theory. Instead, I imagine the virtue theorist's response to be a simple act of contraposition: the example refutes transmission.

Against this response I can revisit the importance of and rationale for extending justification through inference. Chapter 1 did not impose conditions of adequacy lightly. Short of that, it might be said that this is not a case in which it makes sense to draw the inference. I believe Jones only because the situation is one in which no question as to his sincerity arises. As the question does not arise I do not form the belief that answers it. I do not entertain and so do not believe R, and there is nothing to disagree about the justification of. If the question of Jones's sincerity does arise, then whatever raises it defeats my justification for P or Q; again there is nothing to dispute.

It just is not *natural*, in context, for a belief that causes the alleged trouble to be formed. If I tell time by a clock, do I infer that the clock is accurate? I know nothing and need know nothing of the clock's reliability to have a justified belief as to the time by reading it. Thus ignorant, I am amiss to form the belief that it is accurate and would not do so, even if I recognize that its (momentary, at least) accuracy is entailed by what I already believe.

Of course, I agree that the imagined inferences are odd; my diagnosis in Chapter 5 of one's reluctance to exercise the license to infer applies to them. However, I do not see that their oddity solves the problem. The transmission principles are noncoercive. I am not obligated to believe by inference; I am not obligated to infer. But nor can anything stop me; that's the point of the principles. The proposed rejoinder amounts to suspending one's license to infer, to disallowing inference where some further condition is not satisfied. But then inference itself cannot be justificatory; it is really this further condition that carries the epistemic weight. For with the further condition one does not need inference. If I am positioned to assess the clock's reliability, then it does not matter that I already have a justified belief as to the time it now reads. I do not have to have checked its current reading to pronounce it accurate. If I am Jones's close friend, any belief I happen to have about his present condition is incidental to the justification of my beliefs about his character. Either we respect the transmission principles, as I require, or we deny them. It is simply an evasion to accede to them selectively.

Still, if I grant the peculiarity of the troublesome inferences, do I not grant that there is a telling objection to transmission, whatever is also to be said in its favor? Well, intellectual life is imperfect; we cannot have everything. I find it a lot easier to assimilate this peculiarity, to let the sense of it dispel with familiarity, than to do epistemology without logic. If conditions are stipulated to be such that it is sensible to take at face value what a clock reads, then it is not all that unnatural to believe not only what it reads but that its reading is accurate. If, to my knowledge, Jones has not recently dined, has no obvious motive to misinform, is not seeking to gain an advantage of sympathy in a competition for the lone remaining succulent morsel of foie gras, is not under pressure from an indulgent grandmother, then sure, I can justifiedly believe that in telling me he is hungry he speaks sincerely. Again, if under normal conditions with respect to mundane matters one could not take people at their word, life would be very different.

9.7 Inferred Reliability

But is there not then a further problem?[20] What if I assemble an extended body of data regarding a clock's accuracy, all by inference from beliefs its readings justify about the time? From these data I infer the clock's general reliability, and CJC justifies the result. I begin without any basis for assessing the clock's reliability; I have justified beliefs only as to the time and as to what the clock in fact reads. By inference I end up with the justified belief that the clock is reliable, without any further information as to the condition or workings of the clock. Maybe I end up an astute judge of character, just by taking people at their word! Sounds like magic.

I first point out that reliability is not just (nor even) accuracy. I can no more infer a method's counterfactual sensitivity to falsity simply from a record of veracity than I can infer a causal hypothesis from a simple correlation. I also point out that CJC cannot really establish even general accuracy. This requires an enumerative induction from instances of accuracy. Although I contend that ampliative inference must be capable of transmitting justification, I have not attempted to extend my theory to ampliation. I do not have a theory of, do not (claim to) know, how justification

---

[20] Here I wish to take account of Jonathan Vogel's (2000) contention that the implications I am drawing from the transmission principles make reliabilism unacceptable.

attenuates under ampliation, nor what forms of ampliation, under what conditions, transmit justification. But I deny that the theory I do not have includes a transmission principle for (straight) enumerative induction. Such a principle would quickly lead to paradox.

This reply is inadequate, however. Although CJC cannot establish the clock's reliability or even its general accuracy, it can establish the property of having been accurate over an extended, closed interval. The attribution of this property simply by inference remains objectionable. I expect that this problem is serious enough to incite some erstwhile reliabilists to contextualism, or other forms of retreat from transmission. My response is that the problem is all the more reason to insist upon the distinction between the justifications of a believer and a belief.

The belief that the clock has been accurate *is* justified, but it is not justifiedly believed absent reason to believe that it was reliably formed. The problem of formation lies not in CJC, nor more generally in transmission, but in the believer's original justification for believing the clock's individual readings. Supposing the clock reliable, checking its readings under conditions ordinary for the operation of clocks justifies beliefs about the time, but the believer is not justified (not on my theory) in holding these beliefs without some reason to believe the clock reliable. This reason, whatever it is, accumulates to underwrite the more general assessment of accuracy via CJC.

The reason needed may not be onerous. That it is a clock, that it agrees with obvious independent indications of the time, that the progression of its readings over a (conveniently) small interval conform to subjective time, that others rely on it; perhaps these are enough to get justification off the ground. Justification comes in degrees, and the justifiedness of the casual observer leaves room for significant contrast with the epistemic position of the technician who verifies the clock's workings. Similarly, it does not require much for me to be justified in believing, hearing him say so, that Jones is hungry, and, on that account, in believing that he speaks sincerely. Someone who knows Jones well, or who possesses more information as to Jones's recent dietary habits, may be better justified.

There are also peculiar cases in which a proposition that cannot be believed justifiedly is entailed by propositions that seem to be. I am not justified in believing that my lottery ticket will lose, for if I were then by epistemic symmetry and CJC I would be justified in believing that all the tickets will lose, which I know to be false (and which cannot in any case be justified, by CJC and a condition of adequacy). Therefore, I am not justified in believing that I will be unable to endow a chair in philosophy next week, for if I were I could by inference justifiedly believe that my ticket will lose.

This limitation does not generalize as far as some philosophers (Hawthorne, 2004) think. I am justified in believing that I will be home this evening, despite the implication that my house has not burned down. For I am also justified in believing the latter. As urged in Chapter 6, there is no lottery for burning houses, only a statistical probability that does not guarantee that any house will burn. If I learn that some houses have burned (and nothing further to identify which), then I am no longer justified in believing that I will be home this evening. But I am justified in believing that I will *probably* be home, and I think that the justifiedness of this belief is enough to dispel the worry that skepticism is the price of transmission.

I acquiesce in the remaining limitation by reflecting that the very point of purchasing a lottery ticket is to create a possibility that one's fortunes will improve. Why buy a ticket unless, thereby, one's normal resignation to deprivation is relieved? However slight this relief, it reverberates throughout one's belief system, converting justified beliefs (and believings) that propositions are true into justified beliefs (and believings) that they are (highly) probable. Some consequences of justificatory inference take getting used to, but it's not that hard.

Inference *is* justificatory. So, whatever the consequences, virtue theory is unacceptable.

Chapter 10
Counterexamples

10.1 Challenges to Necessity

Is the reliability of the process by which a belief is formed necessary for the belief's justification? Are there not justified beliefs that would have been formed though false, by the method that did form them, even under normal conditions? Consider a person who mistakes certain bushes for trees, but is a reliable identifier of redwood trees. Observing a redwood, he believes justifiedly that there is a tree before him and this belief is justified, but he might have held this belief falsely by observing a bush.[1]

The example is suggestive, but it lacks sufficient plausible detail to determine that the subject's belief is unaffected under the hypothetical supposition of its falsity. More generally, claims about what a subject would believe were his actually true belief to be false, appear arbitrary or stipulative without further information to identify the situation (that would be) responsible for the belief's falsity. This is to be expected from the discussion of subjunctive conditions in Chapter 7. A belief can go wrong in all sorts of ways. An effective challenge to the necessity of reliability for justification must tell us enough about how the belief goes wrong to enable us to decide what this way of going wrong makes the subject believe.

It cannot tell us these things without first rejecting skeptical scenarios. If skepticism is in play, it is wide open what would make one's belief false, and the detail we need can then be withheld compatibly with the attribution of virtually any belief to the subject. Why, in the absence of a redwood, would it be the presence of a bush that makes the subject continue to believe there to be a tree? Why not the hallucination of a redwood? Why not suppose that although the subject continues to know a redwood when he sees one, he no longer—as a result of some intervention—knows them to be trees? Now he thinks they are bushes, so that the hallucination of a redwood does not impart the belief that there is a tree.

The possibility of skeptical scenarios disallows the sort of contextual constraints or presuppositions necessary to determine what is true under the conditions responsible for the falsity of the belief. It is then indeterminate what, if anything,

---

[1] This example is adapted from Alvin Goldman (1976).

J. Leplin, *A Theory of Epistemic Justification*, Philosophical Studies Series 112,
DOI 10.1007/978-1-4020-9567-2_10, © Springer Science+Business Media B.V. 2009

the subject *does* believe under these conditions. That what is believed truly would have been believed falsely becomes mere stipulation. Why not posit something that preserves the truth of the subject's belief, like the presence of an oak tree that the subject is manipulated into mistaking for a redwood? Having rejected skepticism, I take common knowledge for granted. I claim entitlement to assumptions as to what, in a hypothesized situation, is true *instead* of what the subject believes. I shall exercise this license in considering potential counterexamples to the necessity of reliability for justification.

First, let me give the redwood case some detail. The subject does know a redwood when he sees one, and he knows them to be trees. He will believe there is a tree before him when there is only a bush (of a certain kind that he mistakes for trees), but not that there is a redwood tree before him when there is only a bush. Suppose he believes there is a redwood, and *thereby* that there is a tree. Well, *this* is not a belief he would have acquired were it false. He must be looking at a redwood. He doesn't mistake bushes for redwoods; no bush would make him believe there is a tree as he comes to this belief in his present situation. If this belief were false, in this situation, he would not have formed it.

What if growing right behind the redwood is the sort of bush he mistakes for a tree? If the redwood weren't there, would the subject still believe, now mistakenly, that there is a tree because he would see the bush instead? Redwoods grow densely. If the one he sees weren't there, he'd see another; indeed, you have to be up close to see just one redwood (and then you see only part of it). His belief would still be true. To make the belief that there is a tree false we have to situate the subject very differently, such that he does not form it via any belief about the presence of redwoods. But then, his reliability with respect to redwoods does nothing to justify the belief, and we get no counterexample to the requirement that a justificatory method be counterfactually sensitive to falsity.

According to my theory, a belief is justified if, under normal conditions, it would not have been formed, by the method the believer intentionally uses, had it been false. The theory does not say that it *could* not have been formed. What *would* occur, as opposed to what could occur, necessarily introduces a context;[2] what would be believed if P were false depends on what else is true in this context. Why, in the case of the falsity of the belief that there is a tree, there would instead be a bush, so that our botanically challenged observer nevertheless believes that there is a tree, is obscure. This is a case in which those whom the grip of skepticism makes reluctant to assume the relevant constraints can imagine all sorts of $\sim P$ possibilities. Lots of these do not matter to the application of my theory, because they do not affect what *would* be true were P false. As explained in Chapter 7, only if we are given enough information about the context to determine (some of) this can we use my theory to decide whether P is justified. In telling us what justification depends on, my theory does not tell us what in fact is justified absent sufficient contextual information. If

---

[2] As "could" is conditioned by a range of modal constraints, I suppose that a judgment of what could occur also presumes a context; but it is certainly a wider one.

you want to know what is justified you need more than a theory of what justification is, just as if you want to know how a material body will move you need more than the laws of motion.

Another complaint about the example is that the supposition that P continues to be believed though false is insufficient to generate a counterexample to the necessity of my reliability condition for justification. For this supposition may change either the method by which P comes to be believed or the normalcy of the conditions under which reliability requires the method to be counterfactually sensitivity to P's falsity. I will not make an issue of normalcy in the redwood example. But it is not clear whether classifying something as a redwood is the same method of forming the belief that it is a tree as classifying it under another concept that one mistakenly takes to pick out a type of tree. Nor is the role of inference clear in the example. Do we have justificatory inference from justified beliefs in one case versus unjustified beliefs in the other? Is any inference in fact occurring? It is also unclear what kind of justification is in question. Although tree-beliefs generated by bush-sightings are unreliable, the subject could have good reason to believe them reliable. He could be justified in holding these beliefs although they are unjustified. The example might then be defused by granting that redwood-generated tree-beliefs are unjustified—the method is unreliable after all, because it misidentifies bushes—but maintaining that they are justifiedly held in parity with bush-generated tree-beliefs.

To pursue these and further avenues of response, let me construct a more plausibly detailed and less stipulative example.[3] The coach of the basketball team (the only person on campus with his own reserved parking space) owns two cars, an Accord and an Infiniti. I know an Accord when I see one, and know it to be a Honda. But I get the makers of Japanese luxury car lines mixed up. I think that Honda makes the Infiniti, and that the Acura is made by Toyota. I espy the coach's parking space and see his Accord, believing thereby that (not just an Accord but) a Honda is parked there. Were this belief false, this would be because the coach had driven his other car today, which he would have parked in the same place and which I would also take to be a Honda. Evidently my belief that there is a Honda is justified even though I would still believe this were it false. [4]

This example, like the former, exploits the fact that one can have partial mastery of a concept, sufficient for making reliable identifications across a proper subset of its extension. In the complement of this subset, things go wrong in a systematic way. It is possible to deny that *any* applications of a concept are justified if mastery of it is only partial. There is some intuitiveness to the claim that I do not *really* know what a Honda is if I take an Infiniti to be one. But this response is not promising.

---

[3] This is not to deny that we could fix the tree example (picking a tree that does not grow densely), but I prefer to work from a better example.

[4] This example is (loosely) suggested by one Fred Dretske (1981) gives in criticism of David Armstrong's (1973) version of reliabilism.

For one thing, partial mastery, of the specified kind, may be the norm, especially if the concept is deep. Concepts are not acquired all at once, and it may be that the justifiedness of some applications is needed to build on to extend one's mastery. The competing view that ascriptions of mastery apply only to the whole, and grow gradually in accuracy as concepts are deployed with greater skill across a greater range, is one I reject for reasons given in Chapter 9. I take this emergentist doctrine to be false to the phenomena; it misdescribes what people can do and how they learn to do what they do. The child who has learned the multiplication tables up through 6 has partial mastery. Repetition and memory have imparted understanding. (The new methods of instruction fail.) The child has justified beliefs as to the products of numbers below 7. He needs these beliefs to build on to extend his mastery. He could not have learned multiplication in reverse, starting with 12. That would have been memorization without comprehension. In developing sophisticated concepts there is a reason why we start with simple cases. The reason is not only accessibility, as though the cost of starting elsewhere were but excess effort. We have to understand simple applications to grasp others. Understanding of multiplication does not dawn gradually across the number system as a whole.

For another thing, it is unclear that I have to know what a Honda (really) is to know one when I see one. In the philosophy of mind, it is argued that consciousness cannot, even in principle, be ascribed to machines because we do not even understand consciousness in ourselves. As we have no idea what makes *anything* conscious, we cannot be justified in deciding that a machine is conscious.[5] But if this (alleged) ignorance does not prevent us from deciding that *we* are conscious, why should it prevent deciding that a machine is conscious?[6] Why, to know that something is conscious, is it necessary to know what consciousness is? For that matter, we certainly can know (I naturally maintain) that a belief is justified without (yet) knowing what justification is. There are troublesome cases, but partial understanding of a concept is surely enough to justify some applications.

I raised the possibility of faulting the redwood example for failure to specify exactly what method of belief-formation is applied. Let us ask, in the car example, whether the method is just looking, or is looking plus inferring. If inference is included, then the method is not the same when the car is an Infiniti as when it is an Accord. In the former case, the method is unreliable; because the inference involves reasoning through a false step, it is not truth-preserving. Admittedly, the introduction of inferential structure opens the possibility of a similar defect in the latter method. What if I believe Accords to be Civics and know that Civics are

---

[5] I find this reasoning in Colin McGinn (1991, Chapter 8; 1999, Chapter 6). In (1999), pp. 189–192, McGinn argues that to know that a machine is conscious requires knowing what consciousness is. It is not enough that it be true that the machine is conscious and that there be (fallible) behavioral indication of this truth.

[6] The answer cannot be that there is introspection in our case. Introspection does not enable us to know that we are conscious; it only enables one to know oneself to be conscious. I assume knowledge of other minds; otherwise, why make machines the issue?

Hondas? I come to believe that the car before me is a Honda via inferring that it is a Civic. In this case my mistake does not seem to pre-empt justification, but neither does it violate the counterfactual standard of sensitivity to falsity. It is another of those cases in which spurious beliefs about what makes a method reliable are to be tolerated.

Distinguishing methods is an attractive solution. But it depends on assimilating believing *P in virtue* of believing *Q* to believing *P by inference* from *Q*. These seem distinguishable. I have portrayed inference as a psychological transition in thought that, at least in some cases, the believer may desist from or arrest at his volition. Believing *P* in virtue of believing *Q* does not seem to have this (much) structure. Familiar with the Accord, I just *see* that the car is a Honda, although my belief that it is depends upon my identification of it as an Accord. This sounds right, but then I do *not* just see that the Infiniti is a Honda, for it isn't. And yet, unless inference is involved, the Infiniti case is exactly similar as to my belief-formation process. My belief that the car is a Honda in this case depends upon my (true) belief that it is an Infiniti and my (false) belief that these are Hondas. So the method in the former case cannot be just seeing either. *Just* seeing is factive and infallible. What I do to identify cars does not have these properties. My method must be something more complicated that includes judgment by appearance.

This suggests a solution that distinguishes the methods without implausibly encumbering the believer with an act of inference. The distinct methods are to identify the object as one (that is, as belonging to one) of distinct types, which are subsumed under a common category; the methods are individuated by type. Subsumption under a type is not inference. One believes the car to be a Honda, not by identifying it as a Honda as such, but by identifying it as a type of Honda. Identification as a different type is a different way to form the belief. The belief that the car is a Honda is now justified, and were it false it would not have been formed by the same method.

Because of its popularity and influence, it is worth noting that a type of example used by Williamson (2000, Chapter 7) against a sensitivity condition for knowledge succumbs to the same analysis. Suppose I see an object very much less than *x* meters in height. Williamson says that I can know *P*: the object is less than *x* meters high, even if, were the object slightly higher than *x m.*, I would still believe *P* because I tend slightly to underestimate. But how do I form the belief *P*? If I believe *P by seeing* that the object is well under *x m.*, then with the object above *x m.* I would not form *P* in the same way. And if I do not form *P* by seeing that the object is well under *x m.*—if my belief *P* is not responsive to the latitude for error that the situation affords me—then, as I underestimate, it is not clear why the belief should count as known or justified to begin with.

Is this solution defensible? Remember that a justificatory method is applied intentionally. It would have to be part of what the believer intends that the car be identified by type, as a condition for justifiedly believing it to be a Honda. I am not sure whether this is realistic. As the intention in this case is presumably unconscious, I am further unsure how to decide the matter. This may be the sort of situation in which a philosopher claims whatever his theory requires of him, free of the danger that competing claims will admit of independent adjudication.

10.2 Unintended Methods

The alternative, of course, anticipated in Chapter 4, is to abandon the aspiration that reliabilism can be a complete theory of justification. My theory does not claim completeness; the conditions, as I have noted, claim sufficiency, not necessity. Difficulties about the justification of necessary truths, about the reliability of mathematical reasoning, and about special cases like the *cogito*, already counsel caution as to the scope of the theory. To these we might add the justification of applications of partially mastered concepts, where one is systematically deceived about the unmastered part. By backing off on necessity, we open the way to investments of credence that are justificatory without involving the intentional use of a method.

The traditional reliabilist is likely to demand this opening. His interest is not to understand epistemic success in science, but to understand epistemic success *at all*. He wants to refute the skeptic, or, failing that, to understand what knowledge and justification are such that the unrefuted possibility of skeptical scenarios does not pre-empt them. His paradigm for justification, therefore, is the kind of belief that it *takes* skeptical scenarios to challenge: belief presumptively formed by ordinary sense perception under favorable conditions uncomplicated by trickery. This paradigm need involve nothing conscious, deliberate, reflective, voluntary, or—I imagine him adding—intentional.

I think that the addition is just that, not a condition extractable from the former suppositions. Coming to believe by seeing is intentional, at least when it conforms to one's general interests and general inclinations to credulity. There is a general *willingness* to believe what one plainly sees to be the case that underwrites the attribution of intentions on specific occasions, independently of occurrent mental states. But what of an occasion that opposes one's interests and inclinations? One might *not want* to see, but be forced to. One might be forced to watch or listen to something particularly obnoxious or repellent. (One favors opera, whereas one's spouse is partial to Broadway musicals.) I will not try to deny that in such cases (and worse) sense perception delivers justified beliefs without intentional application.

Further to the point, one may have beliefs about one's mental states without doing anything intentional to get them. Labeling one's formation of beliefs about one's mental states "introspection" suggests that there is a method to follow. But in some cases, at least, one just has the belief without any discernable process of coming to believe. Stubbing one's toe, one both is in pain and believes that one is without doing anything, at least not intentionally.[7]

---

[7] Does this mean that the belief can be implanted without the agency of the believer? There does not appear to be agency here, but neither is there (direct) implantation. The belief remains a *response* to the sensation, which alone might be implanted.

Memory might also require an opening for justification without intention. One certainly remembers things without intending to remember them, without intending to remember at all. Perhaps, however, to sustain a belief by memory is justificatory only if the original acquisition of the belief was reliable. Here again, I am reluctant to legislate. Intuitively, memory can be justificatory even in cases where no access to the conditions of a belief's acquisition is possible. But perhaps where lost acquisition was unreliable, the memory is merely apparent. What one genuinely remembers is not that P but only that one believes that P. Still, there are cases in which the strength of a memory seems itself to be evidence that the belief was reliably formed. One retains no memory of its formation, but seems always to have known it. It may be difficult to account for the strength of the belief within one's belief-system without attributing truth. A theorem of mathematics that one cannot (any longer, if ever) prove could be like this. It must be true because I remember it to be a theorem. My inclination is to deny that strength of (apparent) memory, or the need to attribute truth to account for it, justifies belief; at most it justifies believing. It is reason to believe that one would not have come to believe what one (apparently) remembers if it were false.

On balance, I find it prematurely concessive to abandon the prospect that reliability can be necessary for the justification of at least a restricted class of beliefs. Many theories of knowledge officially restrict their scope to the empirical. And while the notion of what is empirical remains vague—sometimes incorporating and sometimes excluding theoretical propositions about experientially inaccessible portions of the natural world—a tenable circumscription of beliefs for which reliability is the correct and complete standard of justification is yet possible. It is evident that neither good reasons nor knowledge require the intentional application of a method of belief-formation. But one can have good reasons to believe without believing, and so without justification. And even if knowledge requires belief, the beliefs it requires could be of a sort that lack justification only because they do not need it.

I will not, however, make an issue of the potential completeness of my theory. At a minimum, we are left with a sufficient condition for epistemic justification. It is not a reductive analysis but it is a theory, still a significant achievement. After all, there is no reason *a priori* to expect epistemic justification to be all of a kind. Epistemic justification advances the epistemic goal, and there could be irreducibly different ways of doing this. A theory of justification, such as mine, can then be correct but incomplete, in that it correctly identifies only some ways in which the epistemic goal is advanced. A pluralism of justificatory conditions is not counterintuitive. If necessity fails, we should see this failure both as a substantial concession on the part of reliabilism and as an important discovery about the complexity of justification.

10.3 Challenges to Sufficiency

Counterexamples to the sufficiency of my reliability condition for justification are more serious. If these arise, the theory requires modification. What about inferences that falsify their conclusions? Suppose that, cowed by an excessively harsh logic instructor, one believes oneself incapable of performing truth-preserving inference. From this belief one infers, evidently without realizing it, that one will perform no such inference today. Assume that if a method of belief-formation is such that its very use falsifies the belief it forms, then it does not justify this belief. Then one's belief that one will not perform a truth-preserving inference today is not inferentially justified.

Could not the belief one infers from nevertheless be justified, perhaps by authority, so that the inferred belief is reliably formed?[8] I think not, for this authority certainly seems unreliable. There is still a potential counterexample, however. One might have good reason to believe the (supposed) authority reliable, and so justifiedly believe oneself incapable of truth-preserving inference.[9] But for this to generate a counterexample we must suppose that while justifiedly believing that one cannot infer, one nevertheless does infer and, moreover, does so intentionally. I submit that this performance defeats the justification one supposedly has to begin with, so that there is no (longer any) justification for one's truth-preserving inference to transmit.

It is crucial to the example that one's initial justifiedly held belief is only a *general* belief in one's incapacity to infer correctly. If in so believing one believes oneself incapable *today*, then the more specific belief cannot be formed by inference; it is present already. The justifiedness of holding the general belief is difficult to square with the intentional use of an inferential method to reach recognition of the more specific incapacity. The believer now has good reason to distrust the general belief, even if we imagine (somehow) that he sustains this belief and is unaware of what he is intentionally doing. Furthermore, unless we suppose that the belief in one's incapacity, rather than just the holding of this belief, is justified, a counterexample will depend on justifiedly believing that the inferential method one is using is reliable. One cannot satisfy this condition while continuing to believe oneself incapable of using this very method.

Consider an opposite scenario, an inference whose very performance requires the truth of the belief it delivers. There is famine in Ethiopia. The Israeli government rescues Ethiopian Jews by transporting them to Israel.[10] An Ethiopian Jewish infant grows up an orphan in Israel, never learning of his foreign origin. He believes he is Jewish, because he assumes he is Israeli born. Although plausible, this is not a very good reason; many Israeli citizens of Israeli birth are not Jews. His belief, we may suppose, is unjustified. Yet, if the belief were false he would never have acquired it. If he were not Jewish, he would have been left in Ethiopia to starve. (We need not assume that he dies, only that he has no occasion to form a belief to infer his

---

[8] This suggestion comes from Ram Neta, in conversation.

[9] This is a concession to strengthen the objection. It is hard to imagine how authority could justify believing that one will never correctly infer. An alternative response to the objection is to argue that to be incapable of inference one would have to be incapable of comprehending one's incapacity.

[10] This much is historically accurate, and suggested the example. A roughly similar example, with an ironic twist, appears in Nozick (1981), p. 185n, to a different purpose.

Jewishness from.) His method of belief-formation is inferential, but the inference is not truth-preserving and its premise is (both false and presumably) unjustified. Yet it might be claimed that my condition for justification of the belief that he is Jewish is satisfied. The method would not have led him to hold this belief if it were false, because if the belief were false the method *could not have been used.*

I propose the qualification that the sufficiency of reliability for justification assumes that the method of belief-formation remains applicable under the counterfactual supposition that the belief formed is false. This assumption, now explicit, has so far been implicit. It is natural in explicating reliability, for if a method is inapplicable, we cannot say what would be believed by its use. To determine what, if anything, the method *would* produce, we must assume that the condition that the belief it *does* produce be false does not, by itself, pre-empt the method's applicability. If it does, we cannot use the theory to judge the belief's justifiedness.

One criticism of reliabilism, remember, was that narrowly enough described, one's method of coming to believe P includes P. Then the truth of P guarantees the method's reliability. Those who do not know how to block the indefinitely extendable specificity of methods cannot grant that one's method remains applicable under the counterfactual supposition that the belief it produces is false, for the belief's truth becomes a defining property of the method. Because of its intentionality condition, my theory is not subject to this criticism. But neither can my theory simply presuppose that with P false the method of believing P remains applicable. Unless it does, the method's reliability, in the sense of counterfactual sensitivity to falsity, is not necessarily justificatory.

But it can be. Let us not overreact. Reliability without counterfactual applicability is not sufficient for justification, but counterfactual applicability is not necessary for justification either. Suppose my method is ordinary vision checked for coherence and so forth, my visual faculty is unimpaired, conditions for viewing are ideal, and the object of sight is readily and unproblematically visible. I believe, correctly, that the object is red, because that's what I see. However, the mad scientist is standing by, ready to intervene should I direct my gaze at anything of another color. He would make me see red, that is, would induce the visual impression of red, were red not visible in the normal way. If what falls within my visual field is not red, then my visual faculty is taken over, compromised so that *I* cannot use it (the scientist is in charge), although I am unaware of the difference. Of course, my victimization is selective and temporary; I'd catch on pretty fast if everything looked red. But while subject to the scientist's intervention, red is all I can see.

Some philosophers will deny that I can see even that.[11] If seeing requires that one's imagery be counterfactually sensitive to changes in the properties of their cause, then I don't see red unless I am allowed to see green (say) as well. I think it is bad enough that I can't see green, and do not understand why these philosophers

---

[11] For example, David Lewis (1986) and some of his respondents discussed in his postscript to his (1986).

won't even let me see the little that the scientist permits me to see. If the scientist, in all his power and authority, allows me to see the red things, who are these philosophers to forbid it? Just as Kepler got some things right while getting lots of things wrong, himself unable to tell the difference, so I see some things that are red and mistakenly think I see other red things too. I have justified beliefs about the redness of things that are red, although were these beliefs false the process by which I form them would not have been available to me. If I am to be philosophized out of my epistemic entitlement to beliefs that things are red, so is Kepler philosophized out of his epistemic entitlement to make what have proved to be among the most crucial advances in the history of astronomy. An epistemology that disrepects this entitlement is not responsive to the data.

If it seems that the scientist's intervention does not pre-empt vision (even) in the nonred cases, if it seems that I *can* see nonred things, just not their (true) color, then try a variation on the example. Suppose that when I look at something nonred the scientist blinds me. Then I form no belief at all, and am not misled about the color of nonred things.

There is a wide range of visual acuity for the sighted. Some people see things that I cannot see. I do not think we want to say that I do not see if, in the absence of what I take myself to be seeing, there would be nothing present visible to me. The fact that I would be unable to use vision to form beliefs were the belief I form false does not prevent its being vision that forms this belief, and so does not prevent this belief from being justified. Seeing does not require that one's imagery be counterfactually sensitive to changes in the properties of their cause.

10.4 Reliably Believed Necessities

Chapter 9 proposed an exception for necessary truths. I prefer not to rely on my theory to explicate the justification of belief in logical necessities, because I do not know in general what would be true if a logical necessity were false (unless it's everything, which is unhelpful). Some specific consequences seem straightforward. If the sum of 16 and 17 were not 33, then when I mix two groups, one of 16 things and one of 17, the size of the resulting group would not be 33. I would count 16 and 17 respectively, but not 33 when I count the mixture. If 9 were prime, it would not be divisible by 3.[12] If water were not H_2O (which mostly it isn't, in fact) it would nevertheless contain hydrogen. If Clark Kent were not Superman, he would be unable to fly.[13] I do not know how to suppose coherently

---

[12] But these consequences may not be as straightforward as they seem. Maybe if $16 + 17 \neq 33$, then I miscount; maybe if 9 were prime then $9 = 3$. How can we tell?

[13] But then, there may not be fictional necessities. Perhaps a consistent continuation of the storyline could reveal Superman to have been masquerading as Kent, a different person, to thwart an enemy. This is not quite the same as a continuation of chemistry revealing that water is something else, for in this case current science is mistaken.

that there is no actual world, but I can say that if there were no actual world, no method of forming the belief that there is would be usable. I take it that the existence of *an* actual world is logically necessary. The belief that there is an actual world is presumably justified, but not on the basis of the reliability of its formation.

However, that it is not possible to suppose counterfactually that a belief is false does not prevent the belief from being reliably produced. If M is a reliable method and conditions are normal for M, then M does not yield falsehoods. So if P were false, M would not yield P. We can draw this conclusion without regard to P's modal status. It is simply a consequence of M's reliability. What we cannot do is *test* M's reliability against propositions whose falsity is not coherently entertainable. But reliability can be established independently. Logical necessities can be entered in encyclopedias and attested to by experts. They can be justified by methods that justify believing contingencies.

The reliability of a method *restricted* to logical necessities, a method, if there is one, inapplicable outside the class of necessities, might not be assessable. If the method yields nothing but necessary truths it trivially satisfies the condition for reliability, but how could this be determined? Checking the method for accuracy against an independent source of information as to the truth-values of its products does not establish counterfactual sensitivity to falsity.

We must distinguish, however, between a logically necessary belief, and a belief to the effect that this belief is logically necessary. One can believe a logical necessity without believing it to be a logical necessity. One can believe a theorem without believing it to be a theorem, and can believe only theorems to be logically necessary. The falsity of the belief that P is a theorem, or is provable, or is necessary, may be entertainable even if the falsity of P is not. In some systems of modal logic, the attribution of necessity to a necessity is contingent, and so its falsity is entertainable. Thus, mathematical and logical methods of reasoning, even if restricted to necessities, can be reliable methods of forming beliefs.

10.5 Second-Order Beliefs

What about my belief that a belief of mine is reliable? If this belief were false, my method of forming it would still be applicable. I form the belief that a belief is reliable by testing the method of its formation. Checking my watch against a standard time signal, I come to believe that my time-beliefs are reliably formed. If I believe this incorrectly, then my watch just happens to read correctly at the time of the signal and is otherwise inaccurate, or the signal itself is inaccurate. In either case, my method, checking against a signal, remains applicable. Of course, if the signal is inaccurate the reliability of the belief that it is accurate cannot be tested successfully against this very signal. But no belief in the signal's accuracy is at issue. It is unclear that such a belief need be (expressly) formed to use the signal to test the reliability of beliefs about the time formed by reading the watch. If I believe the signal is accurate but it isn't, the method used in coming to believe the signal accurate—presumably comparison with another signal—remains applicable.

There are, nevertheless, limitations to the use of my theory to justify beliefs of the second order. The theory does not justify the belief that there are methods of forming beliefs, because if this belief is false then there is no method by which it is formed. This belief is entailed by the belief that there are reliable methods of forming beliefs, which in turn is entailed by the belief that a particular method is reliable. These entailments can be a basis of inference only if what is inferred is true. So my theory does not apply to them. Truth-preserving inference from reliable beliefs is reliable, but its reliability is justificatory only on the condition that the falsity of what is inferred not pre-empt performing the inference.

It may be questioned whether even this concession is sufficient. Reconsider my Moorean example from Chapter 9 of a proposition that could be true but if true could not be believed. This is a second-order proposition P about a proposition Q, to the effect that I believe Q falsely; $P: B(Q)\&\sim Q$. Notice the qualification that P be true. $\sim P$ is strictly compatible with $B(P)$ because $B(B(Q))$ does not imply $B(Q)$; as I have urged, one can be mistaken about what one believes. But of course, what it is plausible for me to believe in this case is $\sim P$. What if *this* belief is false, so that $B(\sim P)$ but P? Then I do believe Q falsely. In particular, $B(Q)$. But if $B(Q)$ then surely $B(\sim(B(Q)\&\sim Q))$. So if my belief, $\sim P$, were false I would nevertheless believe it.

My question, however, is by what process or method I form or sustain this belief. In justifying believing any conjunction of what I justifiedly believe, CJC does not justify believing that I believe nothing falsely. This would require believing that my believing something is sufficient for its truth, and I do not believe that. For that matter, CJC does not justify believing that I believe anything justifiedly; it justifies only on the condition that justification is already present. Nor is $B(\sim(B(Q)\&\sim Q))$ obtainable by any straightforward, truth-preserving inference from what I believe. A second-order belief about what I do or do not believe is not inferable from the semantic content of my first-order belief. We appear to have an example of a proposition that I can believe justifiedly although were it false I would (likely) believe it anyway. But the example does not require any further qualification of my theory. It threatens theories that require the simple sensitivity condition that what one justifiedly believes one would not believe were it false. But these theories fail anyway, as I argued in Chapter 5. A theory that relativizes justification to method is unthreatened.

In an example from Robert K. Shope (1984), what I believe is P: It is true of some of my beliefs about beliefs that I might not have had them. If $\sim P$, then I nevertheless believe P because P is one of those beliefs that I then must have. I agree that P can be justified and justifiedly believed, but only via a method that is inapplicable on the supposition $\sim P$. Justification in this case requires registering the impermanence of second-order beliefs about beliefs, which impermanence cannot be registered if $\sim P$.

I admit that my theory's delivery on its promise to guarantee the transmission of justification through truth-preserving inference is incomplete at the second order of belief. Certainly I admit that the theory does not give a complete account of justification at this level; I have not claimed it to do so even at the first level. I maintain that truth-preserving inference from justified beliefs is justificatory independently of reliability, but this is an additional thesis.

The condition that no methods are reliable does not prevent one from using a method that leads one to believe that some are. But it might be thought that any method of forming the belief that some methods are reliable will presuppose the truth of this belief. For, to assess the reliability of any method would seem to require presupposing the reliability of another. I will respond to this worry in Chapter 11. For now, I grant that trying to establish reliability reliably is like trying to use sense perception to establish its own veridicality. The belief that some methods are reliable may not be counterfactually sensitive to falsity, because if it is false then a presupposition of the use of a method to form it is unsatisfied.

But the belief that there are reliable methods may be reliably formed by inference from the reliability of a particular method, for this inference *is* counterfactually sensitive to falsity. If there are no reliable methods, then the belief that there are cannot be inferred from reliably formed beliefs. This inference, further, is justificatory, for it remains performable on the supposition that there are unreliable methods. That is, the reliability of the beliefs from which one infers is not a condition for performing the inference.

This may seem slender compensation. Ultimately, the justification of second-order beliefs about the reliability of methods would seem to depend on what sources of justification other than the reliability of methods emerge under a pluralistic regime. But if there *are* reliable methods, then, whether or not the belief that there are is reliable, there are justified beliefs.

10.6 Self-Guaranteeing Belief

What if a belief, though contingent, is such that it *could* not be formed if false, regardless of how it is formed in fact? This is presumably a stronger condition; if it could not be formed then it would not be. Consider the case of a self-fulfilling belief, one that could not be believed if false, by any method or none, because believing makes it true. Perhaps the first example to come to mind is the belief that there are beliefs. If this is false then it cannot be formed, because if it is false *no* belief can be formed. Again, my theory does not apply, because no matter how I form this belief, if it is false then my method is inapplicable. As urged in Chapter 3, the existence of an applicable method requires at least the possibility that it issue beliefs; this is not possible under the counterfactual supposition that there are no beliefs.

To challenge my theory, we need a case where the method remains applicable, but is irrelevant because the belief itself, regardless of its mode of formation, guarantees its own truth. A child covets a toy and comes to believe, unjustifiedly, that he will receive it for Christmas.[14] Maybe his friends are promised the toy, and this makes

---

[14] This example originates in one of the many examples used to test epistemological theory that, like success, have multiple authors.

him expect it too. The method is to see what one's friends are getting and then, if they are getting something in common, to believe one will get the same thing. (There is not much room for variety in Santa's sack.) We stipulate that this method is not justificatory. The child's mother learns of his expectation, and, not wishing to disappoint, comes through with the gift. If the belief is false, the child must not come to hold it, for his holding it is causally sufficient for its truth. So the only way for the belief to be false is for the friends not to receive this gift (or not to advertise their expectation). This does not render the method inapplicable; it just prevents the method from yielding the belief. We seem to have reliability sufficient for justification.

I take this to be a case of singular causation. There is no nomic regularity between wanting things and getting them, not even between a child's wanting things and his mother getting them, that the case instantiates. Accordingly, we can perfectly well entertain the possibility that although the child communicates his expectation effectively to the mother, she fails to deliver. So the method is not reliable. Can we construct a case in which the relation of the belief to its truth is nomic?[15]

Suppose the belief that P is physical state α, which a causal law connects to physical state β, and the semantic content of P is that β occurs. I believe P because, in distress, I consulted a psychic who assured me that my condition is not emotional but physical; specifically, it is due to β (for which he happens to market a remedy). If P were false then I could not believe it, but presumably I could still use my method. The psychic would have to be selling something else.

There are suppositions here that may not be realizable. I suggested earlier that an adequate material theory of mind cannot allow beliefs to be discrete physical states, especially not states that have physical states as intentional objects. Here I add that the attribution of truth-value to a physical state is not merely counterintuitive, but paradoxical. Suppose P is physical state α, whereas the semantic content of P is that one is *not* in α. Then P is a belief one holds if and only if it is false. Or suppose, to avoid self-reference, that P is state α and P's semantic content is that Q is true, where Q is a state β that causally prevents α. Then P is believed only if false. If β is also causally necessary to prevent α, then P is believed if and only if P is false. These results are paradoxical in that there is paradox in the inevitability of error no matter what one's doxastic state with respect to a proposition.

---

[15] There could be an intermediate case in which the relation is not nomic but still stronger than that effected by the intervention of the mother. Maybe the mad scientist can be enlisted to ensure that my beliefs are true. If he does this by arranging the world to fit my beliefs, the case most immediately challenges the sufficiency of a safety condition for justification; safety so achieved does not seem justificatory. However, in line with my view in Chapter 8, I think the case can be treated equivalently as a challenge to sensitivity. The scientist is ensuring that I do not hold false beliefs, regardless of my method. Such a case succumbs to the same analysis I will give for the case in which the relation is nomic. Alternatively, one can simply entertain a scenario in which the scientist is distracted, disabled, on vacation, or changes his mind. I see no principled way to rule such a scenario distant from or irrelevant to a world in which he continuously operates.

But I think the difficulty can be defused without speculation as to constraints on the physical nature of belief. Trusting the psychic is not a reliable method, because it would lead me to believe all sorts of things even if they are false. By hypothesis, if P is false then if the method is used it must deliver some *other* belief. Under the counterfactual condition $\sim P$, the psychic *does* tell me something (or he is no psychic). What the method cannot do is lead me to believe P if P is false. For believing P puts me in a state that makes P true. Believing the psychic *when he says that P* is a reliable method.

But, first, it is a method that it is wildly implausible to imagine anyone intentionally using. How could believing by this method be what one intends? People who consult psychics do not condition their credulity on the specific content of the counsel they receive, not unless they are willing to trust the psychic only about what they have independent reason to believe. And then trusting the psychic is not their method of forming (any) beliefs. And, second, given what P is in this case, the method of believing the psychic when he says that P, like any method whose application yields P and P alone, *is justificatory*. The resulting beliefs are justified, but no one is justified in believing them. So I submit that the intentionality requirement of my theory obviates the problem. But should you insist that believing the psychic when he says that P could be one's intended method, then my distinction between justified belief and believing justifiedly accommodates the intuition that the method cannot be justificatory.

This case resembles clairvoyance. We have hit upon a method that happens to be inerrant, but that we have every reason not to credit. I have argued that a method of forming beliefs cannot have its output definitionally incorporated into it, and this present (supposed) method is less plausible than clairvoyance in its unmotivatable restrictiveness to a single proposition. However, it is more plausible than clairvoyance in supplying an understandable scenario for belief-formation.

Upon consideration, I hold to the sufficiency of reliability as a standard of justification, modulo the counterfactual applicability of the method.

10.7 Ranges of Reliability

It may be suspected, however, that to achieve sufficiency for justification, the standard for reliability has been set unreasonably high. We have encountered, intermittently and in diverse guises, the prospect that an otherwise reliable method may be pressed into service beyond its capacities, and, so deployed, may yield false beliefs despite the normality of conditions. Sense perception could produce a false belief under conditions favorable for sight, if the sighted object is camouflaged. In judging the presence and properties of surrounding objects visually, one presumes the absence of trick set-ups, of fake exteriors, of holographic imagery. But what if the object is *naturally* camouflaged, evolved to be *mistakenly* identified in its natural environment by normally perceptive predators? In identifying persons by appearance, one assumes that one's acquaintances have not withheld the information that they have identical twins, nor do they employ doubles to show up in their place. But appearance could produce mistaken identifications without such machination if relied upon where acquaintance is slight. A reliable reference work will not presume to pronounce as to the facts on unsettled issues that authorities contest; it proclaims not *P*, but only *P* according to a particular tradition, person, theory, or line of argument. The latter proclamations should be safe, but mistakes are still possible if the issue is difficult, complex, and specialized.

According to the *Encyclopedia of Philosophy*'s article on the "History of Epistemology", Karl Popper thought that truth is an illusion.[16] I disagree. Since Popper certainly does not consider falsity an illusion, it's hard to see how he can have thought truth to be. The illusion, for Popper, is only to suppose theoretical truth certifiable through empirical enquiry. In fact, Popper writes (1956, p. xxxi), "I uphold the ancient theory of truth according to which truth is the agreement with the facts of what is being asserted." Automatically to attribute all mistakes like this misrepresentation of Popper to abnormal conditions would seem to beg the question as to the reliability of the method.

Here is a contrived example. A thermometer is a reliable gauge of temperature within its normal range of sensitivity. Suppose one associates temperature with weather. One thinks that when it is warm out it is sunny. One gets false beliefs using a reliable method under normal conditions, because the method's reliability does not extend to the propositions one uses it to judge. This is my diagnosis of the difficulty in all these cases. They purport to be counterexamples, not to my analysis of reliability, but to the claim that there are, in fact, (very many) methods that satisfy the analysis.

Some examples will be deflected by my provision that further or more diligent application of a reliable method will correct whatever errors it produces. I think this takes care of vision; in fact, we only (ultimately) learn that vision does err by taking a closer look. Even without the provision, it is unclear what difficulty vision poses in a case like camouflage. Does it deliver false beliefs or none? If the problem is only that one fails to see what is there before one's eyes, there is no exception to my analysis. I would also invoke the provision in response to cases of misidentification. It is implausible that one persists in a false belief upon close inspection of a person one thought one recognized, supposing that what appearance delivered in the first place was indeed a belief, and not just an impression or inclination. But the encyclopedia resists this solution, as does the thermometer. One does not root out the error by reading further, nor correct one's misuse of the thermometer by scrutinizing its readings.

These latter examples strongly suggest that reliability itself, and not just the efficiency with which a reliable method is used, can vary across the normal range of a method's application. The problem, then, is to identify the subrange within which the use of a method is not justificatory, because false beliefs may be formed under normal conditions. This will be the complement of the range within which

---

[16] Paul Edwards, ed., (1967), Volume 3, p. 37.

the method is inerrant. But, of course, the subrange cannot be *identified* by this condition, or every method becomes trivially reliable: it never yields falsehoods under normal conditions across the range of beliefs it justifies, because this is just its range of inerrancy. The problem is to fashion an independent criterion. This problem has no general solution.

The leading intuition is that, as a point of epistemic propriety, methods have to be used with sensitivity to their limitations. The thermometer is an extreme case of impropriety. One is insensitive to the limitation that thermometers do not indicate weather. To justify a belief by intentional use of a reliable method, one must apply the method to what it is reliable at. This is an empirical question.

If I say that *Paul Bocuse* is reliable and *Joy of Cooking* not, or that CNN is reliable and FOX not, this is because my experience affords comparisons of accuracy. I must be prepared for the possibility that further experience will require more refined comparisons. CNN could turn out to be unreliable on certain matters, as encyclopedias evidently are. CNN could be wrong, not just because of an abnormal condition, but systemically, as the *New York Times* became corrupted not through negligence or inadvertence but in the service of a false ideology of compensatory justice. One nevertheless gets justified beliefs from CNN, and, through it all, from the *Times*, not in virtue of a statistical improbability of error but because their standards are counterfactually sensitive to error across a wide range of matters. A philosophical theory cannot fix this range; indeed, it is *not* fixed but shifting. It is an empirical matter what methods are reliable and what are their domains of reliable application.

10.8 Comparisons of Justifiedness

I cannot say that I am satisfied to leave it at this, but I think I must. The fact that justification varies in degree implies some variation in the sources of justification. Having rejected frequentist, and, more generally, probabilistic interpretations of reliability as unfaithful to the epistemic goal, it behooves me to discover some alternative basis for flexibility. Reliability is absolute, and so, accordingly, is the justifiedness of beliefs when reliability is its source. But a justified belief is the more justified the more centrally it is located within the range of the method's counterfactual sensitivity to error. Although beliefs are justified throughout this range, the difficulty, inconstancy, and defeasibility of circumscriptions of this range suggest that centrality is safer. Most central are a method's paradigmatic applications that provide the data from which to theorize about justification. Applications that extend one's belief-system at the risk of error are peripheral.

Of course, there are other bases for appraising the extent of justifiedness. Justification is improved by independent confirmation of the normality of conditions. That one have positive reason to believe satisfied the presuppositions of the use of one's method is a stronger condition than the condition that one lack reason to distrust them. It should, accordingly, strengthen one's justification, and, moreover, do so in the measure that one's reason is good. Someone who confirms that the clock is ticking has a better justified belief as to the time than someone who merely notes its reading. The belief of someone who checks the clock's interior for dust is better justified still.

Another dimension of appraisal is afforded by the confluence of a variety of independent reliable methods. Justification is strengthened by corroborating one's sources. We do not, to recall Wittgenstein, verify what a newspaper reports by checking further copies of the same newspaper, but by comparing what it says with other papers. Justification is reinforced by the agreement of authorities tested for professionalism, disinterest, and the enforcement of precautions against error.

The obvious standards of normalcy and consensus do not, however, capture the comparison of justifiedness within the range of a single reliable method. Along this dimension, appraisal depends on the location of a belief that a method delivers within the method's range of reliability. If we could fix this range precisely, location within it would not matter. But we believe justifiedly by use of the method without fixing this range precisely. And this ability introduces a distinctive element of contrast in the epistemic status of justified belief. At the second order, we are more secure in the justification of a central belief than of a peripheral one, and this difference produces a difference in the degree of justifiedness of holding them.

The variety of directions we may pursue in strengthening justification carries the potential for conflicting assessments. This is no more than the ambiguity already expected from the prospect that justification itself, irrespective of its degree, issues from a plurality of sources. If there are irreducibly different ways to advance the epistemic goal, it is natural that there be irreducibly different influences upon how well this goal is being advanced.

10.9 The Theory, Again

Bringing together the refinements and qualifications I have settled on, and unpacking the definition of reliability, I can state my theory as follows, for believer S, believed proposition P, and method M:

A. *S's belief that P is epistemically justified* if

1. S forms or sustains P by intentionally applying M.
   and
2. P belongs to a domain D of propositions within which M is subjunctively sensitive to falsity: Under normal conditions, M, thoroughly and skillfully applied, would not lead S to believe a proposition in D if it were false; mistakes by M within D are correctable through further application of M.
   and
3. M remains applicable within D under the supposition that P is false.
   and
4. S has no epistemically justified belief incompatible with P.

B. S is epistemically justified in believing P if

1. *S* has good reason to believe that his belief that *P* is epistemically justified. and

2. *S* has no reason to believe that the conditions under which he has used *M* to form or sustain *P* are abnormal.

If mine were a complete theory, then I would not need the notion of epistemic justification in A4; I would refer instead to incompatible beliefs that satisfy A1-A3. But, as I have explained, I do not claim completeness for the theory. It is insufficient for justification that a belief be compatible with beliefs reliably formed, if other justificatory ways of believing can produce defeaters. The notion of epistemic justification invoked in A4 reverts to my initial explication of epistemic justification in terms of truth-conduciveness.

New to the present formulation of the theory is the relativization of reliability to *D*. *D* is not to be confused with the field *F* (or range, in my version) of propositions that the virtue theorist thinks is the primary object of justification. For the virtue theorist, *P*'s justification derives from its membership in *F*; on my theory, *D* is merely the domain within which the method that justifies *P* is justificatory. On virtue theory, the degenerate case $F = \{P\}$ is disallowed because justification depends on an intellectual faculty or capacity that is individuated by reference to *F.* *F* comes first; to decide whether *P* is justified we ask whether it belongs to a field within which the believer is intellectually virtuous. On my theory the degenerate case $D = \{P\}$ is not disallowed; it is the case of believing the psychic when he tells me *P*. It can be dismissed for some *P* on the basis of *A3*, but it fails to arise in any event because of its incompatibility with facts about the intentional use of methods.

In applying my theory, methods are identified by the beliefs they deliver, not by the ranges within which they are reliable. The specification of *P* in a given case determines *M* in that case, and *D* depends on the reliability of *M* thus determined. It is possible on my theory for the scope of one's justified credence to be indefinitely narrow. Centrality demands width. The narrower the scope, the less purchase the *more justified* relation will have—in one dimension, anyway—within one's belief system. The deficiency of nascent belief-systems is not to be ineligible for justification, but to be relatively impoverished in respect of epistemic discrimination.

Chapter 11
Intuition and Method

11.1 Projecting Reliability

Having a developed, defended theory of epistemic justification before us, it is time to seek perspective on the nature of the enterprise, and locate this contribution to it within the wider context of epistemological theorizing. How do my theory and its provenance compare, not just with rival accounts of justification, but also with classic contributions to the epistemological tradition? In this final chapter I draw out and assess some broader implications of what I have been up to.

To pronounce a method of belief-formation reliable requires information as to the truth-values of beliefs it delivers. Of course this is not a sufficient condition. A judgment of reliability is a *projection* from performance, not a record of it. The basis for projectability, I have argued, is an explanatory connection between the reliability of the method and the truth of one's reasons for judging it reliable. In so arguing, I assumed that the beliefs the method has delivered are justified. I assumed that one does not have to be justified in believing a method reliable *as a condition* for obtaining justified beliefs via the method. The reliability of a method is an explanation of information one independently possesses. How do we get this justificatory information without already having methods to whose reliability we can attest?

Perhaps we need not *attest* to the reliability of these methods; it is enough that they *be* reliable. If they are, and we use them, then we get the information needed to project the reliability of the method in question. But if we have no reason to believe the methods we use reliable, if we have no reason to prefer the methods we in fact use over others that we might have used and that would have delivered a different verdict, then we cannot take ourselves to have established the reliability of anything, even if we have. Even if we are judging reliability reliably, and so are doing exactly what our theory of justification requires of us to judge correctly, we have no justification for our judgments.

In Chapter 10 I pointed out that to assess the reliability of one method would seem to require presupposing the reliability of another. This suggestion poses a potential problem of regress in judgments of reliability. The present problem is one of circularity. It claims that judgments of the truth of beliefs and judgments of the reliability of the methods that deliver them depend reciprocally on one another. It is possible to block the regress by escaping the circle. If information as to the truth of beliefs that a method of belief-formation delivers is available independently of information as to the method's reliability, then an explanatory argument may justify projecting the method's reliability, which can then be assumed in assessing the reliability of other methods.

My own view is that the needed independent information is very much available.[1] I take justified beliefs as data, having rejected the skeptical challenge to the legitimacy of doing so. In particular, I take ordinary perceptual beliefs to be justified, or, at least, to be justifiable upon scrutiny for coherence and for the suitability of the relevant environment. Any remotely plausible explanation of the justification of perceptual beliefs requires that ordinary perception be reliable as a method of forming beliefs.

Still, without retracting the confidence, expressed in Chapter 4, that in paradigmatic cases of justification the believer is justified and so are his beliefs, there is a complication to admit. I cannot be so unqualifiedly sanguine about your belief that you have hands as about my own belief that I do. That is, people can come to believe all kinds of things in all kinds of ways, and in offering hands as a paradigm of justification I do not deny that one's belief in them could be unjustified or be held unjustifiedly. Furthermore, my own justification, though secure, is not immediate, but reflective. I have reason to believe that I come by the belief that I have hands reliably, which reason is constructed by considering alternatives. One who never considers the matter at all may not be justified, although his belief is justified. Rather, the situation may be such that one's own justification is simply not at issue.

So, for example, I have justified beliefs that I have two hands, that I am seated at a desk, and so forth. How do I get this information? I do not mean the information that these beliefs are justified; that, as I said, I take for granted. I mean the information that I justifiedly possess about my hands, my posture, the desk. How do I get information, not about the justification of my beliefs, but about *these objects*? It is not plausible to suppose myself in possession of such information unless sense perception is reliable. I could possess these beliefs though perception were unreliable, but in assuming that these beliefs are justified I assume that I possess not just the beliefs but information about the world. Perception has to be reliable for me to have that. I thereby justify the belief that perception is reliable, whence perception is to be trusted in assessing the reliability of specialized, artifactual methods of inquiry.

I now ask those who reject my view, who take skepticism seriously and so find the problems of regress and circularity compelling, to bracket these problems and attend with me to one more fundamental.

---

[1] So I am, in Roderick Chisholm's terms (1973), a "particularist". The circularity I mean to break Chisholm calls the "problem of the criterion". Chisholm denies that the circularity is breakable. He contends that "we can deal with the problem only by begging the question" (p. 37). This is because skepticism remains for him a viable position.

11.2 Self-Referential Consistency

A theory of justification sets conditions for justified belief. To decide whether these are the *right* conditions, whether the theory gets justification right, we must assume that we have justified beliefs to measure the theory against. But without presupposing some theory's standard of justification, we cannot make this assumption. This, in general form, is the problem of circularity. In raising the problem, one takes for granted that the conditions the theory sets for justification are *satisfiable*, that what they require can be achieved. The question is then how to tell, without begging the question, that they are justificatory. I distrust the underlying assumption, made by all parties to debates over the significance of this circularity, that we *can* do what a theory of justification requires for justification.[2]

Bracketing the problems of regress and circularity, or resolving them as I propose, the assumption that we can do what my theory requires for justification is unproblematic in connection with methods of forming first-order beliefs about the world. We can demonstrate the existence of reliable methods of forming such beliefs. We can verify the reliability of thermometers, clocks, and encyclopedias. I have argued, further, that the belief that a belief is formed reliably can be formed reliably. The belief that a method is reliable can be reliable, because the method of projecting reliability that I have described is counterfactually sensitive to falsity. But the assumption that we can do what my theory requires becomes problematic for methods of forming beliefs as to how first-order beliefs get justified. How do we form or sustain the belief that reliably formed or sustained beliefs are justified? Is this belief reliably formed or sustained? Can we satisfy the conditions for its justification?

This is a question not of circularity, but of self-referential consistency. Assuming that my theory gives a correct account of justification, can it be justified? Does belief in the theory satisfy the standards that the theory itself imposes for justification? Notice that an affirmative answer is at most a necessary condition for the correctness of the theory, or, more strictly, a condition of its adequacy. This is why the correctness of the theory is here an *assumption* (for the sake of argument). A self-referentially consistent theory can easily be false. The theory that the best theory is the one that faces the most problems faces the most problems, and is to be judged false *on that account*. My question is whether my theory, or, by extension, any theory so conceived and so dedicated, can satisfy *even* the insufficient condition of self-referential consistency. Independent of the question of the reliability of second-order beliefs about the reliability of methods, there is the question of how to establish the justificatory status of beliefs that reliable methods deliver.

Reliabilism—essentially the view that reliability is epistemically justificatory, of which my theory is a version—is a philosophical theory developed and defended by philosophical methods. It is not easy to say just what these methods are.

---

[2] Debates over the perniciousness of the circularity have made this assumption without hesitation. See, for example, the exchange between Ernest Sosa and Barry Stroud (1994).

Practitioners are notoriously unsuccessful analysts of their own methods. Do not even try to read artists on art. Scientists on science are not much better. They are realists, pragmatists, instrumentalists, positivists, and social constructionists all at the same time. The philosophy of science tries to say what scientific methods are, and comes to conclusions very different from those scientists themselves endorse. Chapter 1 of most any introductory text in a science (especially a social science) proclaims as unproblematic an account of "the scientific method" that philosophers of science reject as naïve, self-defeating, unhistorical, and possibly incoherent. What happens to philosophy when the methods at issue are philosophical? Are philosophers better positioned to discern their own methods than are scientists to discern theirs? Perhaps so, for what philosophy is is supposed to be a topic within philosophy. But what makes it one, that what *anything* is is a topic within philosophy? Perhaps what makes philosophy philosophical is only that we are unsure about it or disagree. And so it is with trepidation that I proceed.

Philosophical method in abstract, general epistemology of the sort I have been trying to do involves formulating universal theses to solve problems, confronting these theses with potential counterexamples, and rejecting or revising them as necessary to withstand these tests. We do not have to worry about the origination of problems or the genesis of an initial thesis; we may assume a context of inquiry rich in leading ideas. A theory like that I have advanced develops through a process of reflecting on what has gone wrong in previous ideas, how the problems can be fixed, and what new problems then arise. This sounds like the common image of science, and indeed, the popular caricature that so distorts science seems to apply to philosophy pretty well.

One difference, both from science and from its caricature, is that the role of intuition in epistemology, and some other branches of philosophy, is not limited to the genesis of ideas, but pervades method. Intuition in philosophy assumes roles that the caricature of science assigns to experimentation and data collection. It determines whether a purported counterexample is genuine, whether, if so, it can be accommodated, and whether the result of accommodation remains tenable. Intuitions are correctable by theory; conflicts do not automatically favor them. But the decision to reject intuition and retain theory must itself be given some basis in (deeper?) intuition, if it is not to be methodologically otiose. At least the decision to favor theory must be shown on balance to incur the lesser conceptual cost, reckoned in the loss of other intuitions that go with theory. The idea of an equilibrium achieved by balancing intuition and theory misrepresents method. Intuitions in philosophical method have the priority that data have in scientific method: they exist independently of theories and it is the responsibility of theories either to get them right or to show cause why they need not do this. There is no reciprocal requirement that intuition conform to theory.

The growing application to issues in epistemology of methods and results of research in artificial intelligence and cognitive science reflects dissatisfaction with the adequacy of philosophical method on this point; it disputes reliance on intuition. So too does the naturalistic movement generally, which seeks an empirical basis for philosophical theories and reconceives philosophy as a continuation of science.

Naturalized philosophy is deeply distrustful of the truth-conduciveness of intuition and adamant in siding with empirical research when it conflicts with intuition. Naturalized philosophy finds deeper methodological morals in the history of radical changes in intuition than in the scientific revolutions that caused them. In view of these developments, I do not think we can stipulate to the reliability of philosophical method.

Several problems are evident in the role of intuition in philosophical method. Intuitions are not uniform. It is commonplace to reach a point at which philosophers simply have to agree to disagree. I do not think that one can know that one has hands without knowing that what one takes for hands are not mere appearance. I must simply admit that some philosophers disagree, and there may be nothing (more) to be done about it. This situation would not be an acceptable outcome in empirical science, nor in a philosophy continuous with empirical science. It would be a problem, possibly persistent but never to be acquiesced in. Until it is solved, the subject is incomplete.

Also, the epistemic status of *being* intuitive is unclear. Abstract general epistemology is a conceptual study. Intuition is therein probative because it is presumed to be revelatory of our concepts. But I have questioned whether there is in fact a clear concept of distinctively epistemic justification available within natural language. It is not clear that what justifies beliefs can be revealed by explicating extant concepts.

Another problem is the long record of failure to recognize weaknesses in philosophical theories. It might look like every philosophical theory, except one's own, is clearly false, and that sustained consensus in favor of any position is unattainable. But in fact there have been traditions long dominant though deeply flawed. The *a priori* of necessity springs to mind. At any moment there might arise another Gettier to challenge millennia of complacency, if not conviction.[3] Can we say with confidence that if a philosophical theory is false, then, under normal conditions, philosophical method will (even eventually, artfully applied) reveal its falsity and not instead condition us to believe it? It is not as though we can begin with a stock of justified philosophical principles and project the reliability of philosophical method from them, as we begin with a stock of justified ordinary beliefs and project the reliability of methods of belief-formation from them. Indeed, the epistemic priority of ordinary beliefs over philosophical principles was part of my argument against skepticism.

---

[3] As I have mentioned, Edmund Gettier (1963) refuted the thesis that one knows whatever one believes truly and with justification. To do this, he pointed out that a true belief may be justified by inference from a false justified belief. I wish to point out that despite the opposition to my contention that truth-preserving inference transmits justification, no one, so far as I have been able to determine, responded to Gettier by denying justification in his examples. Some philosophers decided that knowledge requires a *stronger form* of justification, but none claimed that the inferred true belief was less justified than the false belief it was inferred from. The response, rather, on all fronts, was to admit the decisiveness of Gettier's refutation. From this reaction, one would think that the refuted thesis had been widely believed. It is perhaps more accurate to say that its flaws had gone unrecognized, and the difficulty of formulating sufficient conditions for knowing were underappreciated.

But if reliance on intuition is problematic or defective as method, what happens to reliabilism as theory? It is difficult to resist concluding that if reliabilism is a correct theory of justification, then it is unjustified; for the method of belief-formation that promotes and sustains reliabilism is unreliable. Either that, or some other theory of justification is *also* correct and the justification of reliabilism depends on this other theory. Reliabilism does not appear justified by its own lights. It is not self-referentially consistent.

11.3 The Competition

But of course, this conclusion discredits reliabilism only as a product of philosophical method; it applies equally to other products of this method. If reliabilism is true these too are unjustified, unless some other standard of justification salvages them. But from the perspective of other philosophical theories, there is no particular reason to expect reliabilism to be true. So this conclusion does not convict other theories of self-referential inconsistency. Unfortunately, however, rival theories of justification prove no better off than reliabilism in this regard. Not only are they unjustified by reliabilist standards; they are unjustified by *their own* standards, just as reliabilism is.

Ironically, the one exception would seem to be the *worst* theory going. Rationalism maintains the existence of an intellectual faculty distinct from perception that authorizes belief.[4] If one is prepared to posit such a faculty, one might as well go ahead and let it authorize itself; what more is there to lose? By contrast, rationalism's traditional opponent, empiricist foundationalism, fails upon self-application. According to foundationalism, a justified belief must either be basic—delivered by experience and not dependent for its justification on other beliefs—or be inferable from basic beliefs. But even supposing (dubiously) that there are basic beliefs, the foundationalist thesis is certainly not one of them, nor is it inferable from them. Nor, to my knowledge, have foundationalists attempted to show otherwise.

Consider coherence theories of justification. According to such a theory, a belief is justified by the conformity of its relations with other beliefs to abstract principles of coherence, consistency, mutual reinforcement (possibly measured probabilistically), and explanation. A justified belief explains, is explained by, supports, is supported by, is consistent with, coheres with, and so forth, what else one believes.

Of course, we can ask what justifies these abstract principles, and accuse the answer, if there is one, of circularity. But I have a further question. I have argued, in Chapter 9, that some abstract standard of coherence governs the very attribution

---

[4] Historically, the beliefs so authorized, according to rationalism, are contingent or synthetic, a justificatory insight into necessities being acceptable to empiricists. Without relying on such distinctions, rationalism may be understood broadly as the thesis that knowledge or justification is to be had independently of experience. Then to say what rationalism is, and not merely what it isn't, some nonempirical faculty of insight or understanding will be have to be positively described and posited.

of belief. But my standard is minimal. It takes inconsistencies and incongruities to subvert belief. So, that the introduction of a standard of coherence into one's belief-system not radically disrupt this system is a condition for the standard's admission to the system as a belief. My coherence standard for belief does not require that this standard itself exhibit coherence relations with other beliefs as a condition for being believed, nor that it in fact be believed, let alone that it be justified. My question for coherence theories is whether the thesis that sustaining coherence relations to what else one believes is justificatory itself sustains coherence relations to what else one believes. Does the thesis that coherence is what justification consists in cohere?

If it does, then why do proponents of coherence theories have to expend so much effort *defending* their theory against objections and counterexamples supplied by intuition? Why doesn't coherence theory explain or get support from, e.g., the plain fact that many justified beliefs do not explain much, are not readily explained, are surprising or incongruous with what we thought we knew, conflict with other beliefs we had or with the justifications we thought we had for them, and so forth? Admittedly, coherence theory tells us how to deal with such dissonance: make the minimal adjustments that restore coherence.[5] But why should we think that this policy will favor coherentism over other theories in our resulting system of beliefs? What if the most coherent system contains the thesis that not coherence but reliability is justificatory? What if the thesis that coherence justifies introduces ineliminable incoherence by clashing with the thesis that coherence is improvable by increasing epistemic risk?

According to coherence theories, no belief is invulnerable; the interest of maintaining overall coherence trumps the credentials of any individual belief. What about the belief in coherence theory itself? Isn't a belief-system more coherent *without* the thesis that coherence and coherence alone is justificatory? I see no reason to expect a coherence theory to cohere better with paradigmatically justified beliefs than do other theories of justification.

But even if it does, and all these problems can ultimately be handled, the real question for coherence theories is why these problems should arise at all. Why isn't it all smooth sailing for coherentism? Is this not what one would expect, upon applying coherentism to itself? Sailing smoothly is the coherentist indicator of correctness. If correct, coherentism ought to mesh effortlessly with the rest of our justified beliefs. Its struggle against opposition, even if successful, is self-discrediting. It is difficult not to conclude that if coherence is the correct standard of justification,

---

[5] The policy of adjusting for coherence subverts the claim of coherence theories to explicate epistemic justification, as I understand epistemic justification. Coherence theorists would like to appeal to the truth of a belief system as the best explanation of its long-term coherence (e.g., Bonjour, 1985). But the believer's implementation of a *policy* of constraining belief and changes of belief so as to achieve and sustain coherence is explanation enough, and does not import truth or truth-conduciveness into the system. I suppose one might yet appeal to truth for an explanation of the believer's *ability to carry out* the policy, in the face of inputs from the environment. But unless this explanation adverts to the independent reliability of inputs, I do not see why truth is needed to explain what the believer achieves.

then coherence theories are *not* justified. At best coherence could be *one* source of justification, and not a source available to coherentism.

In one respect, reliabilism is better off than coherentism. Reliabilism permits logical and other forms of necessity; it allows some beliefs to be invulnerable to revision. It thereby reserves the option of making an exception of itself, by limiting its application to contingent beliefs while classifying philosophical theses, or some subclass of philosophical theses to which it belongs, as necessities. It is then not self-referential, and so not self-referentially inconsistent. But this is progress only if necessities enjoy some special justificatory status, such that the question of accounting for their justification, in the general way that reliabilism purports to account for the justification of contingent beliefs, need not arise.

There is, indeed, a tradition of assimilating necessity with *a priority*, where *a priori* beliefs are propositions known (and so, presumably, justified) independently of (whatever) procedures justify contingent propositions. But this tradition is, for present purposes, defective on two counts. The distinction between necessity and contingency must contend with the (powerful) coherentist objection that it is, at best, contextual and revisable. And in telling us that necessities are justified *a priori*, the tradition tells us only what does *not* justify them (experience), not what does. It leaves us without an account of the justification of reliabilism, or of other philosophical theses arrived at through conceptual analysis.

11.4 Constructive Intuitionism

I am not going to branch off, here, to defend the distinction between necessity and contingency against coherentism, nor to propose a special theory of justification for believing necessary truths. Even if the distinction is absolute, it does not follow that any distinctive mode of justification will be available for beliefs of the necessary kind. If necessity is uncertifiable by methods that justify contingencies, it may be uncertifiable altogether.

Instead, I want to ask what our epistemic attitude should be toward the situation as it now stands. We have a theory of justification that fails to account for its own justification. It is nevertheless consistent with its own justification, either because it applies only to contingencies whereas it itself is necessary, or (more promisingly) because it does not purport to be complete but leaves open the possibility of other justificatory standards. Can this theory not "account" for the justification of lots of beliefs, and so constitute a philosophical advance, even if we are not in a position to rule on the justification of the theory itself?

Well, if the theory is true, then it *does* account for lots of justification, whether or not we think it does. The fact that we *have* the theory, that it has been brought into existence and entertained, then constitutes philosophical progress even if we do not recognize this progress to have occurred.

So suppose that the theory is false. Does this make it worthless? Maybe a *bad* theory, a theory that makes no sense or that there is strong reason to disbelieve, is worthless, or worse. But reliabilism is certainly not a bad theory (so far as I can see). And surely the falsity of a theory is compatible with its representing a significant intellectual advance, on many fronts. It provides *an* answer to a seemingly important question. Even if this answer is incorrect, it gives us (defeasibly) reason to think that the question is answerable; the question is not confused or pointless. In capturing correctly our judgments about an enormous range of data that we hold it responsible for explaining and against which we test it, the theory provides a way of conceptualizing these data that is pragmatically useful even if mistaken. And a false theory, sometimes better than a true one, reveals strengths and weaknesses in other theories and provides a basis for further theorizing. The correct attitude, then, is that reliabilism is progressive whether true or not.

Compare the situation in science. Even if we were not justified in believing scientific theories, we would still be justified in believing that empirically successful theories constitute scientific progress. They tell us more than we knew about (nontheoretical aspects of) the world, and this additional information is justified by their success. Although we do not justifiedly believe them, we do justifiedly believe them to be predictively reliable. We have reason to believe what we use them to tell us that is not reason to believe *them*.

In the worst case, our attitude toward philosophy should be as positive as such a constructive empiricist attitude is toward science.[6] Constructive empiricism about science denies that theoretical beliefs are justifiable, but nevertheless claims justification for the empirical beliefs they deliver by inference. We should hold, as a minimum position, a *constructive intuitionism*, according to which we are justified in believing that intuitively successful epistemological theories, as well as theories in other areas of philosophy that have not (yet) been naturalized, constitute philosophical progress.

Specifically, success in capturing intuitions about the application of concepts analyzed in philosophical theories is projectable. An epistemological theory that correctly reproduces settled judgments as to the justification of first-order beliefs will correctly adjudicate new, difficult, or controversial cases. We can therefore use it to understand more than we could without it, quite independently of the issue of its own justification. We can use it, for example, to dispense with skepticism.

It is greatly to the credit of constructive intuitionism that it refutes constructive empiricism. For in licensing the projection of our justificatory practices, constructive intuitionism assures us that there is no special barrier to the justification of theoretical beliefs in science. We explain the empirical success of scientific theories by crediting them with partial or approximate truth. Certain forms of empirical success have no other explanation. According to the view I have advanced here, this explanatory connection provides good reason to judge reliable our method of investing credence in theories. Using empirical evidence to judge, not just the ability of theories to explain and predict this evidence, but also their truth is a reliable method

---

[6] Constructive empiricism is the antirealist position of Bas van Fraassen in (1980).

of investing credence. Theoretical beliefs that result from its intentional application are therefore beliefs that we are justified in holding.

Newton was justified in his theoretical identification of the phenomenon of weight exhibited in free fall on earth with orbital motions in the heavens. He had good reason to believe that his methods would not have led him to this conclusion were it false. By the second decade of the 20th century, physicists had overwhelmingly good reason to believe in the atomic structure of matter. If matter did not have an atomic structure, they would not have been able to measure Avogadro's number in 16 independent ways. Examples of theoretical beliefs that my theory says scientists are justified in holding are readily available.

Of course, some justified theoretical beliefs turn out to be false. I doubt that my theory accounts for all of the ways in which this can happen, but it does account for some. For example, it is often difficult to verify that the presuppositions of the use of a reliable method are satisfied. In turning out to be false, a reliably formed belief can indicate the importance of a neglected variable. Then too, methods believed for good reason to be reliable can turn out not to be, so that unjustified beliefs were justifiedly held. More generally, the complexity of scientific reasoning in comparison with more ordinary methods of belief-formation suggests that reliable methods in science will require prolonged and repeated application by numerous practitioners to get things right. Scientists get things wrong using reliable methods that justify false beliefs, when their use of these methods is insufficiently scrupulous and diligent. It is often further application of the very method by which a false justified belief was formed that reveals the error.

But let us not exaggerate the problem of reconciling the falsity of justified theoretical beliefs with the reliability of their formation. Widely discussed in the philosophy of science is a "skeptical historical induction" to the falsity of current science from the failures of past theories that once enjoyed broad acceptance based on strong empirical support. I criticized the structure of this induction, in particular its reliance on second-order evidence, in Chapter 6. But even if the induction succeeds, and we conclude that all theories, whatever their evidential support, are or are likely to be false, it does not follow that the justified theoretical beliefs of scientists are false. For the justified theoretical beliefs of scientists are rarely beliefs to the effect that a particular theory is true, as such. The examples above are broader and weaker than unqualified endorsements of particular theories. Newton, after all, was never sanguine about gravity, and denied that he could even construct a theory of gravity. Instead, what the scientist will, typically, believe of a theory is that it is partially true, or, at the strongest, that it is approximately true. Properly explicated, these semantic properties typically *are* possessed by the historical theories whose ultimate failure the skeptical induction analogizes to current science.

Under the conditions I hold sufficient for justified theoretical belief, theories are at least partially true. What the scientist justifiedly believes under these conditions is, then, true simpliciter. According to my version of scientific realism, justified theoretical belief requires a sustained and unblemished record of novel predictive success. As I understand novelty, there is good reason to believe that under normal conditions a theory would not attain such a record if it were not partially true.

So what the scientist justifiedly believes, if his standards for credence are those I defend, is, typically, both true and reliable. Thus, the theory of justification proposed here combines with my theory of scientific realism to advance, if not fully to complete, the project of understanding epistemic justification in science.[7] If we cannot have justification in epistemology, we can yet have it in science. And understanding how we can have justification in science is what (much of) epistemology is for in the first place.

Some philosophers will want more than the constructive intuitionist position can deliver in epistemology. But some will want less. They will want the naturalistic project pushed without limit. They will dispute the legitimacy of any philosophy that this project leaves out. I think this project has considerable, further potential. For example, it ought to extend throughout metaphysics, if metaphysics is to tell us the nature of reality. Intuition cannot decide that, although much contemporary, post-positivistic metaphysics proceeds as though it can. It cannot decide the ontological status of possible worlds, for example. But there is a reason to think that naturalism stops short of obviating conceptual studies in epistemology.

Science does not speak to the distinction between knowledge and justified belief, nor to the distinction between justification and evidential support. A philosophical theory sensitive to these distinctions is not decidable by empirical methods; it is not continuous with science. There may be no scientific basis even for my leading assumption that the belief that one has hands is justified. In fundamental physics there are no principles of individuation or reidentification for hands. There is no fact of the matter as to what particles they contain nor what regions of space-time they occupy. An epistemology based on hands is ultimately responsive to concepts for which intuition is the essential method of analysis. And for such an epistemology, constructive intuitionism is the best I think we can do.

---

[7] My (1997) defends the realist claims made here and explicates the notions of partial truth and novelty that these claims employ.

Bibliography

Alston, William (1995), "How to Think about Reliability", *Philosophical Topics*, vol. 23, pp. 1–29.

Armstrong, David (1973), *Belief, Truth, and Knowledge*, Cambridge University Press, Cambridge.

Austin, John (1962), *Sense and Sensibilia*, Oxford University Press, Oxford.

Bach, Kent (1985), "A Rationale for Reliabilism", *The Monist*, vol. 68, pp. 246–263.

Bergmann, Michael (2006), *Justification without Awareness*, Oxford University Press, Oxford.

Blackburn, Simon (1984), "Knowledge, Truth and Reliability", Henrietta Hertz Lecture, *Proceedings of the British Academy*, vol. 70, pp. 167–87.

Bonjour, Laurence (1985), *The Structure of Empirical Knowledge*, Harvard University Press, Cambridge.

Bonjour, Laurence (2003), "A Version of Internalist Foundationalism", in *Epistemic Justification*, Bonjour, Laurence and Sosa, Ernest, eds., Blackwell Publishers, Malden, MA.

Brueckner, Anthony (1991), "Unfair to Nozick", *Analysis*, vol. 51, no. 1, pp. 61–64.

Chisholm, Roderick (1973), "The Problem of the Criterion", *The Aquinas Lecture*, Marquette University Press, Milwaukee.

Christensen, David (1993), "Skeptical Problems, Semantic Solutions", *Philosophy and Phenomenological Research*, vol. 53, pp. 301–320.

Cohen, Stewart (1999), "Contextualism, Skepticism, and the Structure of Reasons", in *Philosophical Perspectives* vol. 13, supp. to *NOÛS*, James E. Tomberlin, ed., Blackwell Publishers, Oxford. pp. 57–91.

Cohen, Stewart (2005), "Contextualism Defended", in *Contemporary Debates in Epistemology*, Matthias Steup and Ernest Sosa, eds., Blackwell Publishers, Oxford, pp. 56–62.

Conee, Earl (1992), "The Truth Connection", in *Evidentialism*, Earl Conee and Richard Feldman, eds., pp. 242–259, Oxford University Press, Oxford.

Conee, Earl and Feldman, Richard (1998), "The Generality Problem for Reliabilism", *Philosophical Studies*, vol. 89, pp. 1–29.

Craig, Edward (1989), "Nozick and the Sceptic: The Thumbnail Version", *Analysis*, vol. 49, no. 1, pp. 161–162.

David, Marian (2001), "Truth as the Epistemic Goal", in *Knowledge, Truth, and Duty*, Matthias Steup, ed., pp. 151–170, Oxford University Press, New York.

DePaul, Michael (2004), "Value Monism in Epistemology", in *Knowledge, Truth, and Duty*, Matthias Steup, ed., pp. 170–187, Oxford University Press, New York.

DeRose, Keith (1995), "Solving the Skeptical Problem", *The Philosophical Review*, vol. 104, pp. 1–52.

DeRose, Keith (2002), "Assertion, Knowledge, and Context", *The Philosophical Review*, vol. 111, pp. 167–205.

Donnellan, Keith (1966), "Reference and Definite Descriptions", *Philosophical Review*, vol. 75, pp. 281–304.

Dretske, Fred (1970), "Epistemic Operators", *Journal of Philosophy*, vol. 67, pp. 1007–23.

Dretske, Fred (1971), "Conclusive Reasons", *Australasian Journal of Philosophy*, vol. 49, pp. 1–22.

Dretske, Fred (1981), *Knowledge and the Flow of Information*, Blackwell, Oxford.

Dretske, Fred (2005), "The Case Against Closure", in *Contemporary Debates in Epistemology*, Matthias Steup and Ernest Sosa, eds., Blackwell Publishers, Oxford, pp. 13–26.

Edwards, Paul, ed., (1967), *The Encyclopedia of Philosophy*, Collier Macmillan Publishers, New York.

Fantl, Jeremy and McGrath, Matthew (2002), "Evidence, Pragmatics, and Justification", *Philosophical Review*, vol. 111, no. 1, pp. 67–94.

Feldman, Richard (1988), "Having Evidence", in *Evidentialism*, Earl Conee and Richard Feldman, eds., pp. 242–259, Oxford University Press, Oxford, 2004.

Feldman, Richard and Conee, Earl (1985), "Evidentialism", *Philosophical Studies*, vol. 48, pp. 15–34.

Fogelin, Robert (1994), *Pyrrhonian Reflections on Knowledge and Justification*, Oxford University Press, Oxford.

Foley, Richard (1987), *The Theory of Epistemic Rationality*, Harvard University Press, Cambridge.

Foley, Richard (1993), *Working without a Net: A Study of Egocentric Epistemology*, Oxford University Press, New York.

Forbes, Graeme (1984), "Nozick on Skepticism", *Philosophical Quarterly*, vol. 34, no. 134, pp. 43–52.

Gettier, Edmund (1963), "Is Justified true belief Knowledge?", *Analysis*, vol. 26, pp. 144–146.

Ginet, Carl (1985), "*Contra* Reliabilism", *The Monist*, vol. 85, pp. 175–187.

Glymour, Clark (1980), *Theory and Evidence*, Princeton University Press, Princeton.

Goldman, Alvin (1976), "Discrimination and Perceptual Knowledge", *Journal of Philosophy*, vol. 73, pp. 771–91.

Goldman, Alvin (1979), "What is Justified Belief?", in *Justification and Knowledge*, George S. Pappas, ed., D. Reidel, Dordrecht, pp. 1–23.

Goldman, Alvin (1986), *Epistemology and Cognition*, Harvard University Press, Cambridge.

Harman, Gilbert (1973), *Thought*, Princeton University Press, Princeton.

Hawthorne, John (2004), *Knowledge and Lotteries*, Oxford University Press, Oxford.

Heller, Mark (1989), "Relevant Alternatives", *Philosophical Studies*, vol. 55, pp. 23–40.

Heller, Mark (1999a), "The Proper Role for Contextualism in an Anti-Luck Epistemology", in *Philosophical Perspectives* vol. 13, supp. to *NOÛS*, James E. Tomberlin, ed., Blackwell Publishers, Oxford, pp. 115–31.

Heller, Mark (1999b), "Relevant Alternatives and Closure", *Australasian Journal of Philosophy*, vol. 77, pp. 106–208.

Hempel, Carl (1945), "Studies in the Logic of Confirmation", *Mind*, vol. 54, pp. 1–26.

Jeffrey, Richard (1983), *The Logic of Decision*, second Edition, University of Chicago Press, Chicago.

Kelly, Thomas (2003), "Epistemic Rationality as Instrumental Rationality: A Critique", *Philosophy and Phenomenological Research*, vol. 66, no. 3, pp. 612–41.

Klein, Peter (1976), "Knowledge, Causality, and Defeasibility", *Journal of Philosophy*, vol. 73, pp. 792–812.

Klein, Peter (1981), *Certainty: A Refutation of Skepticism*, University of Minnesota Press, Minneapolis.

Klein, Peter (2000), "Contextualism and the Real Nature of Academic Skepticism", *Philosophical Issues*, vol. 10, pp. 108–116.

Kyburg, Henry (1961), *Probability and the Logic of Rational Belief*, Wesleyan University Press, Middletown, CT.

Kyburg, Henry (1970), "Conjunctivitis", in *Induction, Acceptance, and Rational Belief*, M. Swain, ed., D. Reidel, Dordrecht.

Leplin, Jarrett (1997), *A Novel Defense of Scientific Realism*, Oxford University Press, Oxford.

Leplin, Jarrett (2004), "A Theory's Predictive Success Can Warrant Belief in the Unobservable Entities it Postulates", in *Contemporary Debates in the Philosophy of Science*, Christopher Hitchcock, ed., Blackwell Publishers, Malden MA, pp. 117–133.

Lewis, David (1973), *Counterfactuals*, Harvard University Press, Cambridge, MA.

Lewis, David (1986), "Veridical Hallucination and Prosthetic Vision", *Australasian Journal of Philosophy*, vol. 58, pp. 239–49.

Lewis, David (1996), "Elusive Knowledge", *Australasian Journal of Philosophy*, vol. 74, pp. 549–67.

Long, Doug (1992), "The Self-Defeating Character of Skepticism", *Philosophy and Phenomenological Research*, vol. 52, pp. 67–85.

McGinn, Colin (1984), "The Concept of Knowledge", *Midwest Studies in Philosophy*, vol. 9, Peter French, Theodore Uehling, Howard Wettstein, eds., University of Minnesota Press, Minneapolis, MN, pp. 529–555.

McGinn, Colin (1991), *The Problem of Consciousness*, Blackwell Publishers, New York.

McGinn, Colin (1999), *The Mysterious Flame*, Basic Books, New York.

Moore, G.E. (1959), "Proof of the External World", in *Philosophical Papers*, Allen and Unwin, London.

Neta, Ram (2003), "Contextualism and the Problem of the External World", *Philosophy and Phenomenological Research*, vol. 66, no. 1, pp. 1–31.

Nozick, Robert (1981), *Philosophical Explanations*, Oxford University Press, Oxford.

Plantinga, Alvin (1993), *Warrant: The Current Debate*, Oxford University Press, Oxford.

Plantinga, Alvin (1996), "Respondeo" in *Warrant in Contemporary Epistemology*, Jonathan Kvanvig, ed., Rowman and Littlefield, New York.

Pollock, John (1970), "The Structure of Epistemic Justification", *American Philosophical Quarterly*, Monograph Series no. 4, pp. 62–78, Basil Blackwell Publishers, Oxford.

Pollock, John (1983), "Epistemology and Probability", *Synthèse*, vol. 55, pp. 231–52.

Pollock, John (1984), "Reliability and Justified Belief", *Canadian Journal of Philosophy*, vol. 14, no. 1, pp. 103–114.

Pollack, John (1986), *Contemporary Theories of Knowledge*, Roman and Littlefield, Lanham, MD.

Popper, Karl (1956), *Realism and the Aim of Science*, Roman and Littlefield, Totowa, New Jersey.

Pritchard, Duncan (2002), "Radical Skepticism, Epistemological Externalism, and Closure", *Theoria*, vol. 66, pp. 129–162.

Pritchard, Duncan (2005), *Epistemic Luck*, Oxford University Press, Oxford.

Putnam, Hilary (1981), *Reason, Truth, and History*, Cambridge University Press, Cambridge.

Roush, Sherrilyn (2005), *Tracking Truth: Knowledge, Evidence, and Science*, Oxford University Press, Oxford.

Russell, Bertrand (1957), "Mr. Strawson on Referring", *Mind*, vol. 66, pp. 385–9.

Shope, Robert K. (1978), "The Conditional Fallacy in Contemporary Philosophy", *Journal of Philosophy*, vol. 75, no. 8, pp. 397–414.

Shope, Robert K. (1984), "Cognitive Abilities, Conditionals, and Knowledge", *Journal of Philosophy*, vol. 81, no. 1, pp. 29–48.

Sosa, Ernest (1985), "The Coherence of Virtue and the Virtue of Coherence", *Synthèse*, vol. 64, pp. 3–28.

Sosa, Ernest (1991), *Knowledge in Perspective*, Cambridge University Press, Cambridge.

Sosa, Ernest (1993), "Proper Functionalism and Virtue Epistemology", *NOÛS*, vol. 27, pp. 51–65.

Sosa, Ernest (1994), "Philosophical Skepticism and Epistemic Circularity", *Proceedings of the Aristotelian Society*, Supp. Vol. 24, pp. 363–290.

Sosa, Ernest (1999), "How to Defeat Opposition to Moore", in *Philosophical Perspectives*, vol. 13, supp. to *NOÛS*, James E. Tomberlin, ed., Blackwell Publishers, Oxford, pp. 141–155.

Sosa, Ernest (2000), "Skepticism and Contextualism", in *Philosophical Issues*, 10, supp. to *NOÛS*, Ernest Sosa and Enrique Villanueva, eds., Blackwell Publishers, Oxford, pp. 1–19.

Stroud, Barry (1994), "Scepticism, Externalism, and the Goal of Epistemology", *Proceedings of the Aristotelian Society*, Supplementary Volume 24, pp. 291–307.

Van Fraassen, Bas (1980), *The Scientific Image*, The Clarendon Press, Oxford.

Vogel, Jonathan (1990), "Are there Counterexamples to the Closure Principle?", in *Doubting: Contemporary Perspectives on Skepticism*, Ross M. and Ross G. eds., Kluwer Academic Publishers, Dordrecht.

Vogel, Jonathan (2000), "Reliabilism Leveled", *Journal of Philosophy*, vol. 97, no. 11, pp. 602–23.

Warfield, Ted (2004), "When Epistemic Closure does and does not Fail: a Lesson from the History of Epistemology", *Analysis*, vol. 64, pp. 35–41.

Williams, Michael (1991), *Unnatural Doubts; Epistemological Realism and the Basis of Scepticism*, Blackwell Publishers, Oxford.

Williamson, Timothy (2000), *Knowledge and its Limits*, Oxford University Press, Oxford.

Wittgenstein, Ludwig (1958), *Philosophical Investigations*, Blackwell Publishers, Oxford.

Wright, Crispin (1985), "Facts and Certainty", *Proceedings of the British Academy*, pp. 429–72.

Wright, Crispin (1991), "Skepticism and Dreaming: Imploding the Demon", *Mind*, vol. 100, no. 397, pp. 87–117.

Wright, Crispin (2000), "Cogency and Question-Begging: Some Reflections on McKinsey's Paradox and Putnam's Proof", in *Philosophical Issues*, 10, supp. to *NOÛS*, Ernest Sosa and Enrique Villanueva, eds., Oxford: Blackwell Publishers, pp. 140–164.

Index

Printed in the United States
140291LV00005B/14/P